PERGAMON INTERNATIONAL LIBRARY
of Science, Technology, Engineering and Social Studies
The 1000-volume original paperback library in aid of education, industrial training and the enjoyment of leisure
Publisher: Robert Maxwell, M.C.

Internal Combustion Engines

IN 2 VOLUMES

A detailed introduction to the thermodynamics of spark and compression ignition engines, their design and development

VOLUME 2

THERMODYNAMICS AND FLUID MECHANICS SERIES
General Editor: W. A. WOODS

Other Titles of Interest in the Pergamon International Library

BENSON	Advanced Engineering Thermodynamics, 2nd Edition
BRADSHAW	Experimental Fluid Mechanics, 2nd Edition
	An Introduction to Turbulence and its Measurement
DANESHYAR	One Dimensional Compressible Flow
DIXON	Fluid Mechanics: Thermodynamics of Turbomachinery, 3rd Edition in SI/Metric units
	Worked Examples in Turbomachinery (Fluid Mechanics and Thermodynamics)
DUNN & REAY	Heat Pipes, 2nd Edition
GIBBINGS	Thermomechanics
HAYWOOD	Analysis of Engineering Cycles, 2nd Edition (in SI units)
HOLMES	Characteristics of Mechanical Engineering Systems
MORRILL	An Introduction to Equilibrium Thermodynamics
WHITAKER	Elementary Heat Transfer Analysis

Internal Combustion Engines

IN 2 VOLUMES

A detailed introduction to the thermodynamics of spark and compression ignition engines, their design and development

VOLUME 2

ROWLAND S. BENSON
Late Professor of Mechanical Engineering, University of Manchester Institute of Science and Technology

AND

N. D. WHITEHOUSE
Reader in Internal Combustion Engines, University of Manchester Institute of Science and Technology

PERGAMON PRESS

OXFORD · NEW YORK · TORONTO · SYDNEY · PARIS · FRANKFURT

U.K.	Pergamon Press Ltd., Headington Hill Hall, Oxford OX3 0BW, England
U.S.A.	Pergamon Press Inc., Maxwell House, Fairview Park, Elmsford, New York 10523, U.S.A.
CANADA	Pergamon of Canada, Suite 104,150 Consumers Road, Willowdale, Ontario M2J 1P9, Canada
AUSTRALIA	Pergamon Press (Aust.) Pty. Ltd., P.O. Box 544, Potts Point, N.S.W. 2011, Australia
FRANCE	Pergamon Press SARL, 24 rue des Ecoles, 75240 Paris, Cedex 05, France
FEDERAL REPUBLIC OF GERMANY	Pergamon Press GmbH, 6242 Kronberg-Taunus, Pferdstrasse 1, Federal Republic of Germany

First edition 1979

British Library Cataloguing in Publication Data

Benson, Rowland Seider
Internal combustion engines.
Vol. 2
1. Internal combustion engines
I. Title II. Whitehouse, Norman Dan
621.43 TJ785 79-40361

ISBN 0 08 022717 1 hard (Vols 1 & 2 combined)
ISBN 0 08 022718 X flexi (Vol 1)
ISBN 0 08 022720 1 flexi (Vol 2)

In order to make this volume available as economically and as rapidly as possible the typescript has been reproduced in its original form. This method unfortunately has its typographical limitations but it is hoped that they in no way distract the reader.

Printed and bound at William Clowes & Sons Limited
Beccles and London

Preface

The ever present energy crisis and the need for environmental controls has had a major impact on the development of the internal combustion engine. In this development a closer understanding of the thermodynamic processes occurring within the engine is necessary. Both authors have been continuously involved in industry and the universities over the past 30 years in the design, development, research and the teaching of internal combustion engines. The present text represents the fruits of some of their labours. Much of the material is original and some has not been published heretofore. The material has been used in the authors' department in the final year's BSc courses and in the MSc course.

The text has been written as a companion to one of the authors (R.S. Benson's) text in the same series entitled *Advanced Engineering Thermodynamics* (2nd edition). A novel feature in the text is the presentation of FORTRAN listings of two programs for simple cycle calculations—one for a compression ignition engine cycle and the other for a spark ignition engine cycle. Methods are also outlined for more complex cycle calculations of the type which are now normally carried our in design offices. The quantitative material for combustion processes in compression ignition engines and some of the data for spark ignition engines are based on the latest research carried out in the authors' laboratories.

The text is divided into two volumes to suit the convenience of students. The first volume contains material suitable for an undergraduate course in internal combustion engines, whilst the second volume is more relevent to postgraduate courses.

The book is primarily concerned with the thermodynamics of internal combustion engines but inevitably we have included hardware features. Since the successful understanding of the processes in which the engine operates is dependent on experimental work, a section is included on experimental methods which is appended to Volume I although some of the techniques are only used in advanced research establishments.

The authors wish to acknowledge with thanks the help of the numerous research students, research assistants and technical staff in producing the data used in the text. They wish to thank the various publishers and institutions for the reproductions of figures, due acknowledgement of which is given in the appropriate place. They also wish to thank Mrs. M. McDonnell and Mrs. P. Shepherd for typing the draft and Mrs. J.A. Munro for typing the camera ready copy of the text.

Finally, they wish to thank their respective wives and families for their patience and forbearance for the many evenings and weekends spent in preparing the text.

Contents of Volume 2

Contents of Volume 1

Chapter 7

The Gas Exchange Process

Notation

a	speed of sound
A	area
C_p	specific heat at constant pressure
C_v	specific heat at constant volume
e	specific internal energy
h	specific enthalpy
H	enthalpy
L	length
m	mass
N	rotational speed
p	pressure
Q	heat transfer
r	volume or pressure ratio
R	gas constant
S	stroke
t	time
T	temperature
V	volume
w_m	molecular weight
x	a/ac (see equation (7.102))
Z	instantaneous scavenge efficiency
α	crankangle
ρ	density
ϕ	area ratio
λ	scavenge ratio
κ	isentropic index, ratio of specific heats
η_{CH}	charging efficiency
η_{SC}	scavenge efficiency
η_V	volumetric efficiency

General subscripts

a	air
c	cylinder
cv	control volume
exh	exhaust pipe
f	fuel
g	gas
o	stagnation
R	in cylinder at exhaust valve or port opening

7.1 THE GAS EXCHANGE PROCESS IN FOUR-STROKE AND TWO-STROKE CYCLE ENGINES

The period during which the products of combustion are replenished by the fresh charge is called the gas exchange period, and the thermodynamic and gas dynamic processes the gas exchange processes. The duration of the gas exchange period and the processes are different in four-stroke cycle engines from two-stroke cycle engines. The sequence of events in a four-stroke cycle engine take place over approximately two engine strokes, whilst in a two-stroke cycle engine the period is approximately one-third of each of two strokes.

The cycle of events in a four-stroke engine and the cylinder pressures are shown in Figs. 7.1 and 7.2. The exhaust valve opens near the end of the expansion stroke (Fig. 7.1(a)) and the cylinder pressure drops from a to b (Fig. 7.2). In an ideal cycle the cylinder pressure at the end of the stroke (BDC) is equal to the exhaust pressure. This period is normally called the exhaust blowdown period. The piston then travels towards top dead centre displacing the products of combustion that are exhausted through the exhaust valve (Figs. 7.1(b) and 7.2). This is called the exhaust stroke. The pressure in the cylinder is constant in an ideal cycle. Before the piston reaches the end of the exhaust stroke the air inlet valve opens (Fig. 7.1(c)). For a short period both the air inlet valve and the exhaust valves are open. This is called the overlap period (c to d in Fig. 7.2). On the return stroke of the piston the exhaust valve closes (Fig. 7.1(d)) and air (or fuel/air mixture) fills the cylinder. This is called the suction stroke. The pressure drops (d to e in Fig. 7.2) to allow for the gas velocity through the air valve. The piston then returns towards the top dead centre and the air valve closes (Fig. 7.1(f)). The trapped pressure corresponds to f in Fig. 7.2.

The exact timing of the valves varies from engine to engine. Some typical timings are given in Table 7.1.

In the supercharged engine the overlap period is longer than in the non-supercharged engine. This provides an additional air flow, which displaces the products of combustion from the clearance volume and also reduces the exhaust gas temperature and cools the exhaust valve. The excess air is called the scavenge air.

Depending on the engine design there might be two air inlet valves and two exhaust valves, but it is normal practice to have a single air valve and a single exhaust valve. The flow area of the air valve is normally greater than the flow area of the exhaust valve. Later in this chapter we shall discuss methods for computing valve areas. The air valve design may also assist the air motion in the cylinder. If the inlet passage is inclined to

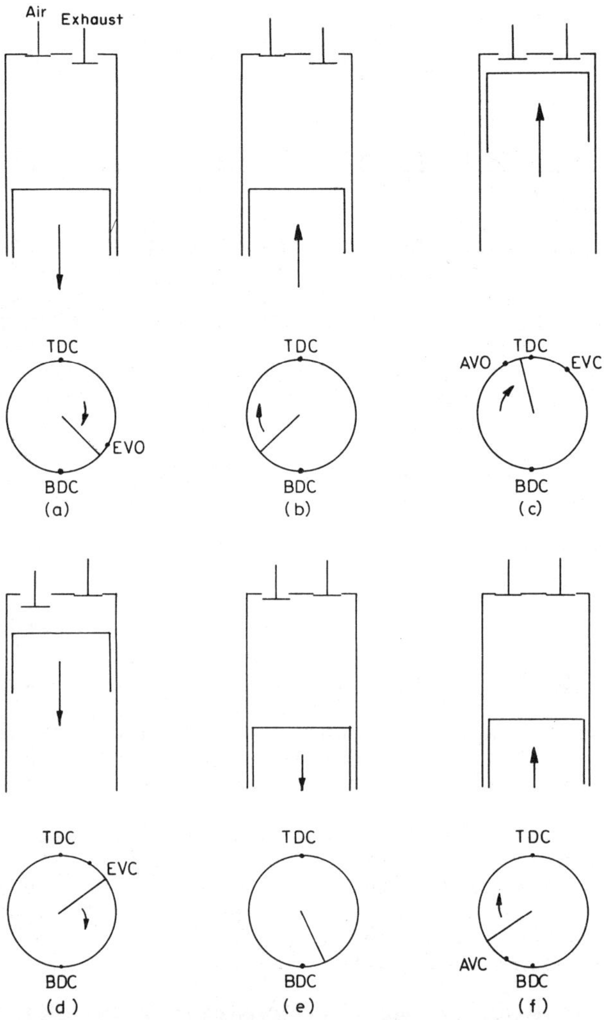

FIG. 7.1. Cycle of events in four-stroke cycle gas exchange process.

the cylinder axis (Fig. 7.3a) or if a mask is fitted in the back of the valve (Fig. 7.3b), swirl is produced in the cylinder during the suction stroke (c to f in Fig. 7.2). This swirl persists during the compression stroke and assists the combustion process.

The turbulence level in the engine cylinder is also influenced by inlet passage and valve design.

The cycle of events in the gas exchange process for a two-stroke cycle engine is shown in Fig. 7.4 with the cylinder pressure diagram in Fig. 7.5. A two-stroke cycle engine may have an exhaust valve or

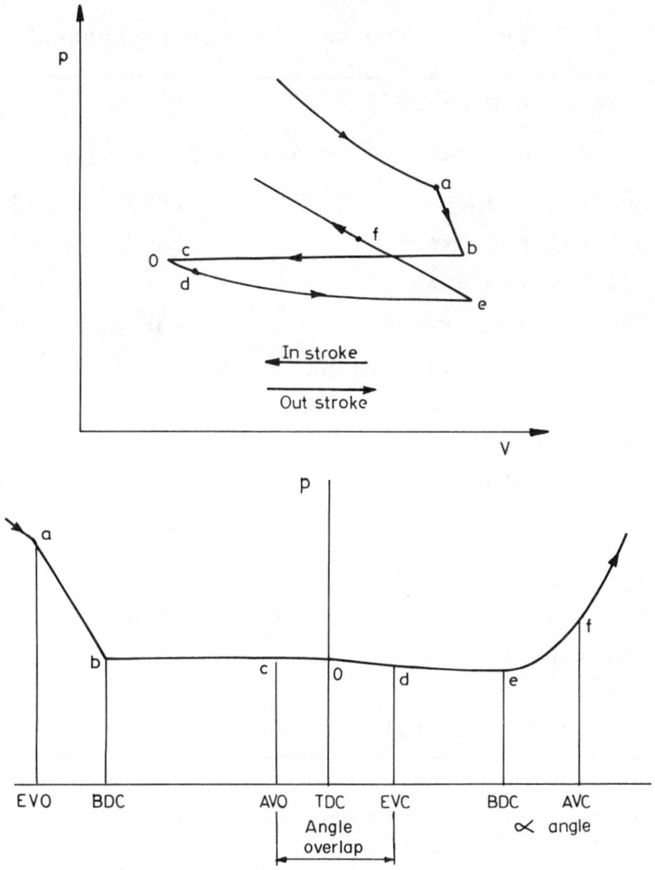

FIG. 7.2. Cylinder pressure diagrams in four-stroke cycle gas exchange process. ab, blowdown; bo, exhaust stroke; cd, valve overlap—both air and exhaust valves open; of, suction stroke.

valves in the cylinder or exhaust ports in the cylinder liner. The air inlet ports are always in the cylinder liner. The two alternative arrangements are shown in Fig. 7.4. After about two-thirds or so of the expansion stroke the exhaust valve or ports are opened (the latter controlled by the piston) and the products of combustion pass to the exhaust. The cylinder pressure drops from a to b (Fig. 7.5). In an ideal cycle when the cylinder pressure equals the air supply pressure, the air ports open (Fig. 7.4(b)). The period from exhaust port or valve open (EPO or EVO) to air port open (APO) is called the exhaust blowdown period. When the air ports and exhaust ports (or valve) are both open the incoming air displaces the products of combustion. This process is carried out by the gas dynamic forces in the cylinder. Thus we see the essential difference between a two-stroke cycle engine and a four-stroke cycle engine, for in the latter engine the piston displaces the products of combustion whilst in the former the air column from the air ports displaces the products. In the cylinder pressure diagram (Fig. 7.5) this period is represented by b to c and is

TABLE 7.1. Typical Timings for Gas Exchange Process.

Four-stroke (Cycle 0-720°)		0° TDC
Non-supercharged (small valve overlap)		
Exhaust valve open	(EVO)	156°
Exhaust valve closed	(EVC)	367°
Air inlet valve open	(AVO)	346°
Air inlet valve closed	(AVC)	582°
Supercharged (with large valve overlap)		0° TDC
Exhaust valve open	(EVO)	142°
Air inlet valve open	(AVO)	287°
Exhaust valve closed	(EVC)	428°
Air inlet valve closed	(AVC)	573°
Two-stroke (Cycle 0-360°)		0° TDC
Exhaust valve (or ports) open	(EPO)	114°
Air inlet ports open	(APO)	136°
Air inlet ports closed	(APC)	224°
Exhaust valve (or ports) closed	(EPC)	246°

called the scavenge period. Depending on the timing, the exhaust valve or ports may close before or after the air ports are closed. If the latter, then period c to d is called the charging period. If the exhaust ports close after the air ports, then some of the charge air will pass into the exhaust. Typical port timings are given in Table 7.1.

There are a number of methods of scavenging. These are illustrated in Fig. 7.6. The arrangement in Fig. 7.6(a) is called cross-scavenging. The air enters the cylinder through ports directed off centre; this produces an upward movement of the air which forces the exhaust gases to travel in the path shown. In the cross-scavenge engine the air port and the exhaust ports are on opposite sides of the cylinder liner. Another arrangement is shown in Fig. 7.6(b), where the exhaust ports are above the air ports. In this case the air stream is directed on to the unported wall and the gas stream follows the path shown. This system is called loop scavenging. A modification of this type of scavenging is the Schnürle system (Fig. 7.6(c)). In this system the ports are on the same level with the exhaust ports located between two air ports. In all the three arrangements of Fig. 7.6(a), (b) and (c), the air path is in a U-direction, in two other systems shown in Fig. 7.6(d) and (e) the air flow is in one direction only. These methods are called uniflow scavenging. In Fig. 7.6(d) the exhaust valves are in the cylinder head, whilst in Fig. 7.6(e) the exhaust ports are at the opposite end of the cylinder to the air ports. In the latter

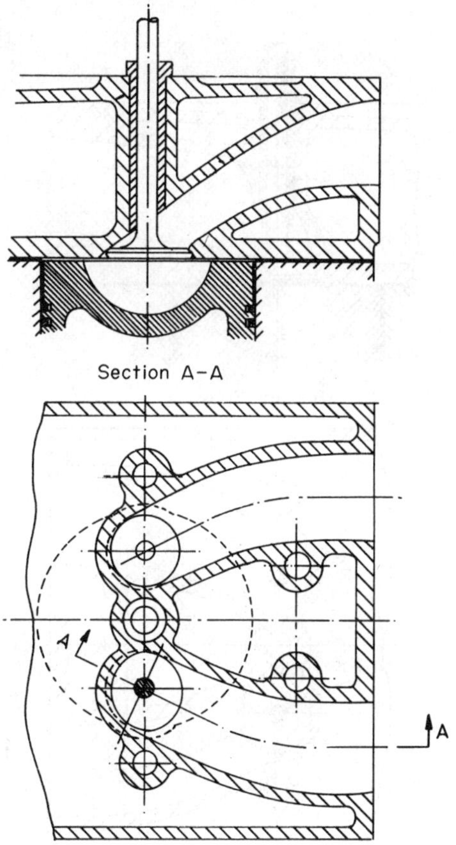

FIG. 7.3(a). Inclined inlet port to product air swirl. (Ricardo and Hempson,[1] by courtesy of Blackie and Son.)

arrangements a second piston is required to control the exhaust port timing. These engines are called opposed piston engines. The two pistons may be linked together to drive a single crank or have separate connecting rod link systems to a single crankshaft or two separate crankshafts geared together. The air ports are normally arranged to produce a controlled swirl as shown in Fig. 7.7. In this case the air motion can cause a central core of the cylinder contents to be not scavenged. We shall return later to discussing the scavenge process in two-stroke cycle engines.

7.2 DEFINITIONS

The efficiency of the gas exchange process is measured in a number of ways. The most efficient process will be that in which the products of combustion are completely replaced with a fresh charge at the charge pressure and temperature. Thus we can define the charging efficiency as the ratio of the mass of air trapped in

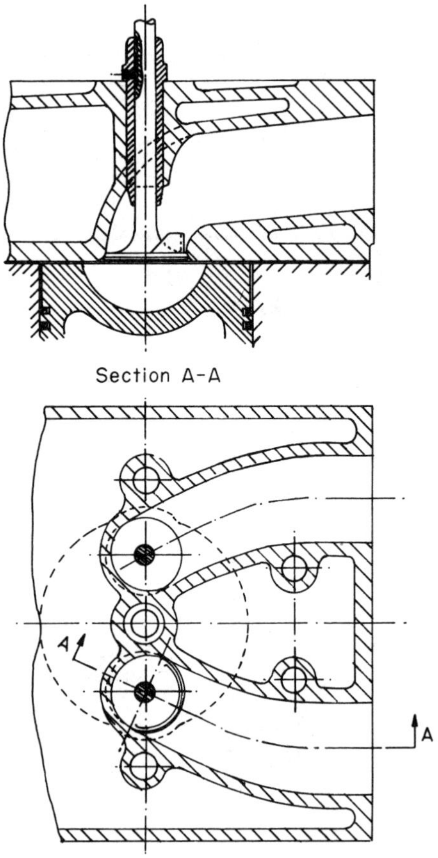

FIG. 7.3(b). Partially masked inlet valve to produce air swirl. (Ricardo and Hempson,[1] by courtesy of Blackie and Son.)

the cylinder to the mass of air which could be trapped if the efficiency were 100%, that is, if the cylinder contained pure air (or fuel/air mixture) only. To measure this efficiency we should require to know the composition of the cylinder contents as well as the trapped cylinder pressure and temperature. To assess the composition of the cylinder contents we use a sampling process. For this purpose we define a scavenge efficiency. This is the ratio of the mass of air trapped in the cylinder to the mass of air plus residuals from the products of combustion trapped in the cylinder. Due to pressure losses across the inlet valves or ports, the effect of heat transfer and the mixing of the residuals with the pure air, the trapped pressure and temperature will be different from the charge pressure and temperature. This can be assessed from the volumetric efficiency which can be defined as the mass of air plus residuals trapped in the cylinder to the mass of air which would be trapped at the supply pressure and temperature.

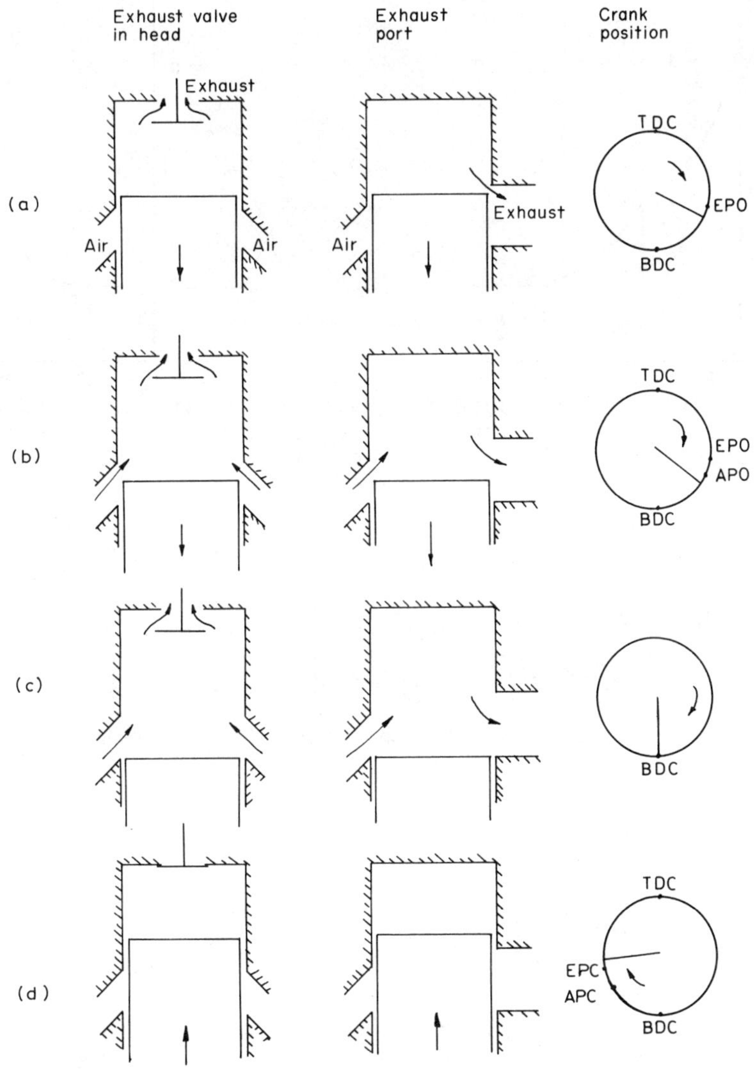

FIG. 7.4. Cycle of events in two-stroke cycle gas exchange process.

In all these definitions the cylinder volume at which the maximum trapped mass of air is determined is the cylinder volume at the commencement of the compression stroke, called the trapped volume. In some texts the reference condition is the cylinder swept volume in place of the trapped volume. For any engine the ratio of the trapped volume to swept volume is a constant.

The ratio of air supplied to mass of air, which could be trapped at the supply pressure and temperature, is called the scavenge ratio. For naturally aspirated engines the ratio will be

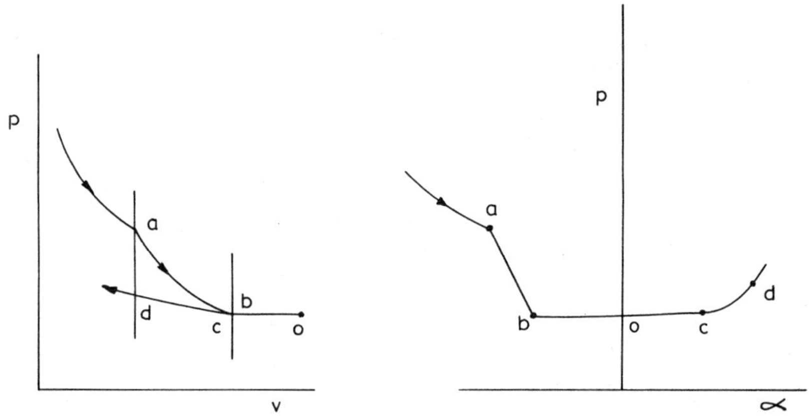

FIG. 7.5. Cylinder pressure diagrams in two-stroke cycle gas exchange process. ab, blowdown; bc, scavenge process—inlet and exhaust ports open; cd, overlap—"charging" extra air if c = EPC and d = APC.

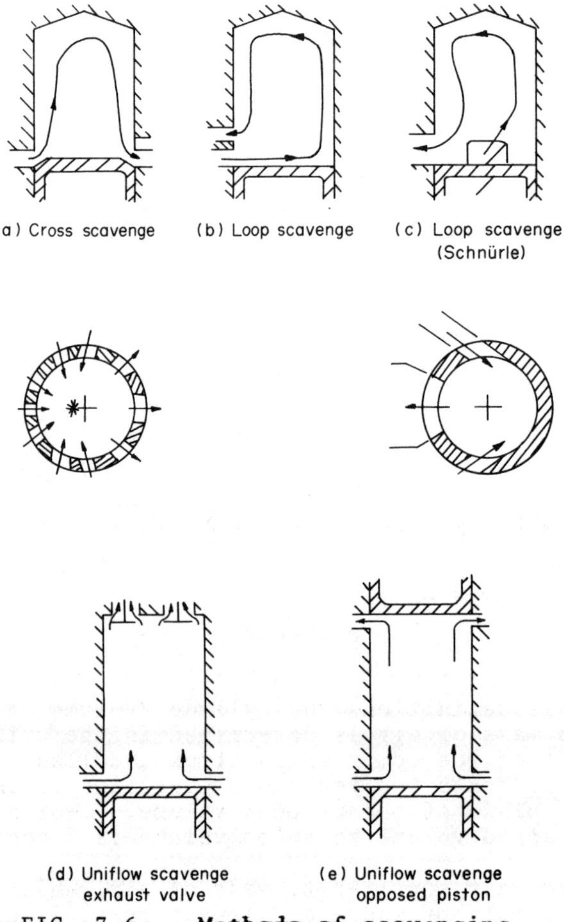

FIG. 7.6. Methods of scavenging.

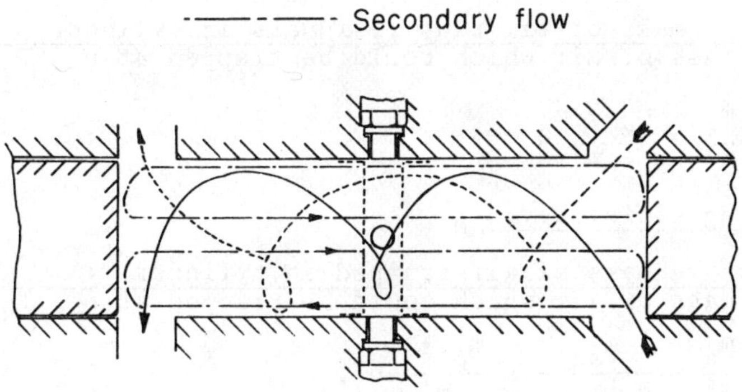

FIG. 7.7. Air motion in opposed piston uniflow scavenged engine.

less than unity. For two-stroke cycle engines with an external air supply and for supercharged four-stroke cycle engines, this ratio is normally greater than unity.

We shall now develop algebraic expressions for the above efficiencies. We use the following nomenclature:

p_s = boost (or scavenge) air pressure

T_s = boost (or scavenge) air temperature

m_a = mass of air in the cylinder

m_g = mass of residuals in the cylinder

m_c = mass of air plus residuals in the cylinder

m' = mass of air which <u>could be</u> trapped in the cylinder at pressure p_s, temperature T_s

m_s = mass of air <u>supplied</u> to the cylinder at $p_s T_s$

We then have:

(a) <u>Scavenge efficiency η_{SC}</u>

$$\eta_{SC} = \frac{\text{mass of air in the cylinder}}{\text{mass of air plus residuals in cylinder}},$$

$$\eta_{SC} = \frac{m_a}{m_a + m_g}. \quad (7.1)$$

(b) Volumetric efficiency η_V

$$\eta_V = \frac{\text{mass of air plus residuals in cylinder}}{\text{mass of air which could be trapped at } p_s,\ T_s},$$

$$\eta_V = \frac{m_c}{m'} = \frac{m_a + m_g}{m'} \tag{7.2}$$

(c) Charging efficiency η_{CH}

$$\eta_{CH} = \frac{\text{mass of air trapped in cylinder}}{\text{mass of air which could be trapped at } p_s,\ T_s}$$

$$\eta_{CH} = \frac{m_a}{m'} = \frac{m_a}{m_a + m_g} \times \frac{m_a + m_g}{m'},$$

$$\eta_{CH} = \eta_{SC}\ \eta_V. \tag{7.3}$$

(d) Scavenge ratio λ

$$\lambda = \frac{\text{mass of air supplied to the cylinder}}{\text{mass of air corresponding to the cylinder volume at } p_s, T_s}$$

$$\lambda = \frac{m_s}{m'}. \tag{7.4}$$

We can re-examine some of these terms using volume rather than mass.

Let V_T be the cylinder volume at the commencement of compression, i.e. the trapped volume, V_a be the volume of air in the cylinder, V_c be the volume of air plus residuals and V_g be the volume of residuals.

Then at cylinder pressure p_c, temperature T_c,

$$m_a = \frac{p_c V_a}{R_a T_c}, \qquad m_g = \frac{p_c V_g}{R_g T_c},$$

and if $R_a = R_g$,

$$\eta_{SC} = \frac{m_a}{m_a + m_g} = \frac{V_a}{V_a + V_g}. \tag{7.5}$$

Thus the scavenge efficiency can be expressed in terms of volume if the analysis of the cylinder contents is made at the same temperature and pressure.

The volumetric efficiency is given by

$$\eta_V = \frac{m_a + m_g}{m'}$$

$$(m_a + m_g) = \frac{p_c \; V_T}{R_c \; T_c} ,$$

$$m' = \frac{p_s \; V_T}{R_a \; T_s} .$$

Now if we assume $R_c = R_a$, then

$$\eta_V = \left(\frac{p_c}{p_s}\right) \left(\frac{T_s}{T_c}\right) . \tag{7.6}$$

Note it is possible to have a volumetric efficiency greater than 100%.

The charging efficiency is given by

$$\begin{aligned} \eta_{CH} &= \eta_{SC} \; \eta_V \\ &= \eta_{SC} \left(\frac{p_c}{p_s}\right) \left(\frac{T_s}{T_c}\right) . \end{aligned} \tag{7.7}$$

For a constant pressure isothermal scavenge process ($T_s = T_c$, $p_c = p_s$), then

$$\eta_{CH} = \eta_{SC} = \frac{V_a}{V_a + V_g} . \tag{7.8}$$

7.3 THERMODYNAMICS OF THE GAS EXCHANGE PROCESS

When the cylinder is open to the inlet system or the exhaust system or both systems, the cyinder conditions are affected by the flow conditions in the inlet and exhaust systems. Furthermore, the composition of the cylinder contents will vary with time. We therefore have to make some simplifying assumptions; these will be indicated in the development of the analysis. We will commence with the exhaust blowdown period and then follow with the exhaust stroke. Next we shall examine the suction stroke and, finally, the scavenge process by a number of methods. The latter process is extremely complex in thermodynamic terms and we must make a number of our simplifications to establish the influence of the air flow on the trapped conditions.

In the analysis which follows we shall assume that the valve and port areas are adequate. Later in this chapter we shall develop methods for computing these areas.

The analysis uses two basic equations, namely, the first law for open systems (equation (2.32)).

$$\frac{dQ}{dt} - \frac{dW_s}{dt} - p\,\frac{dV}{dt} = \frac{\partial(E)_{cv}}{\partial t} + \left(\frac{dm}{dt}\right)_{out} h_{o\ out} - \left(\frac{dm}{dt}\right)_{in} h_{o\ in},$$

where $$h_o = h + \frac{u^2}{2}$$

and the mass continuity (equation (2.33)),

$$\left(\frac{dm}{dt}\right)_{cv} = \left(\frac{dm}{dt}\right)_{in} - \left(\frac{dm}{dt}\right)_{out}$$

7.3.1 Exhaust Blowdown Period

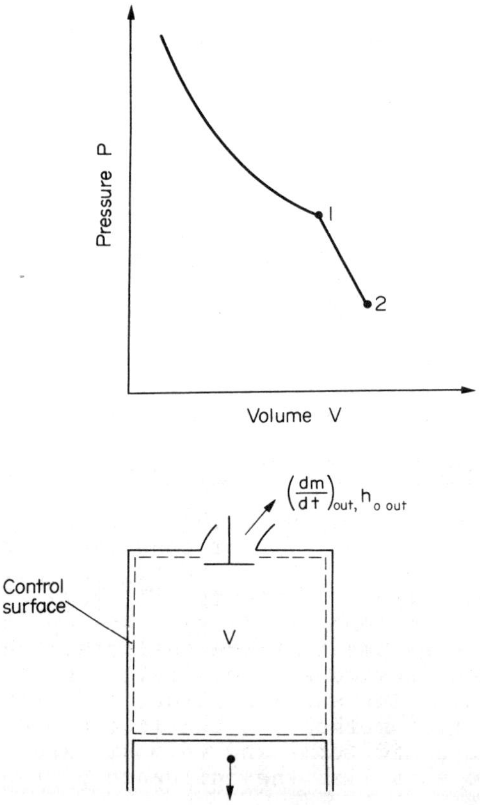

FIG. 7.8. Exhaust blowdown period. m, mass in cylinder; V, volume; e, specific internal energy; h, specific enthalpy.

In Fig. 7.8. the pressure/volume diagram during the blowdown period is shown. This process occurs both in a four-stroke and two-stroke engine. The pressure at valve opening is p_1 and the final pressure is p_2.

Subscript 1 refers to conditions at the beginning of the period and subscript 2 to the end.

We shall assume an adiabatic process; hence $dQ/dt = 0$. No gas flows into the cylinder and $(dm/dt)_{in} = 0$.

For the control system shown the internal energy in the control volume is me. Hence we can write the first law for the system (2.32) as

$$- p \frac{dV}{dt} = \frac{\partial(me)}{\partial t} + \left(\frac{dm}{dt}\right)_{out} h_{o\ out}. \tag{7.9}$$

For a time step dt during the blowdown period, (7.9) becomes

$$- p\ dV = d(me) + h_{o\ out}\ dm_{out}. \tag{7.10}$$

The mass continuity equation for the control system (2.33) is

$$\left(\frac{dm}{dt}\right)_{cv} = - \left(\frac{dm}{dt}\right)_{out}. \tag{7.11}$$

For a time step dt during the blowdown period it follows that

$$dm_{out} = -dm. \tag{7.12}$$

We now assume that the total enthalpy of the gas leaving the cylinder $h_{o\ out}$ is equal to the enthalpy of the gas in the cylinder h.

Thus $$h_{o\ out} = h. \tag{7.13}$$

If we expand (7.10) and substitute for $h_{o\ out}$ from (7.13), we have

$$- p\ dV = m\ de + e\ dm - h\ dm$$

or $$- p\ dV = m\ de - (h-e)dm$$

or $$- \frac{p\ dV}{(h-e)} = \frac{m\ de}{(h-e)} - dm$$

or $$- \frac{p\ dV}{m(h-e)} = \frac{de}{(h-e)} - \frac{dm}{m}. \tag{7.14}$$

We shall assume an ideal gas with constant specific heats, with $e = h = 0$ at $T = 0$; then

$$h-e = \left(C_p - C_v\right)T = RT \tag{7.15}$$

and $$\frac{de}{h-e} = \frac{C_v dT}{RT} = \frac{1}{(\kappa-1)} \frac{dT}{T}. \tag{7.16}$$

Substituting (7.15) and (7.16) into (7.14) we have

$$\frac{dV}{V} = \frac{dT}{(\kappa-1)T} - \frac{dm}{m}\,. \qquad (7.17)$$

For the blowdown period 1→2 the relationship between the properties of the gas is obtained from integration of (7.17); thus

$$-\ln\left(\frac{V_2}{V_1}\right) = \frac{1}{\kappa-1}\ln\left(\frac{T_2}{T_1}\right) - \ln\left(\frac{m_2}{m_1}\right)$$

or

$$\frac{m_2 V_1}{m_1 V_2} = \left(\frac{T_2}{T_1}\right)^{\frac{1}{\kappa-1}} \qquad (7.18)$$

Now from the state equation

$$\frac{m}{V} = \frac{p}{RT}\,.$$

$$\frac{m_2 V_1}{m_1 V_2} = \frac{p_2}{p_1}\,\frac{T_1}{T_2}\,.$$

It follows then that

$$\left(\frac{p_2}{p_1}\right)\left(\frac{T_1}{T_2}\right) = \left(\frac{T_2}{T_1}\right)^{\frac{1}{\kappa-1}}$$

or

$$\frac{p_2}{p_1} = \left(\frac{T_2}{T_1}\right)^{\frac{\kappa}{\kappa-1}} \qquad (7.19)$$

Thus the expansion in the cylinder is isentropic during the blowdown period. The ratio of the mass at the end of the blowdown period m_2 to the mass at the beginning m_1 is

$$\frac{m_2}{m_1} = \left(\frac{V_2}{V_1}\right)\left(\frac{p_2}{p_1}\right)^{\frac{1}{\kappa}}\,. \qquad (7.20)$$

The conclusion that the process is isentropic arises because we have assumed that the stagnation enthalpy leaving the cylinder $h_{o\,out}$ is equal to the cylinder enthalpy h. This in turn implies that the gas stream is flowing from the ports under adiabatic conditions. Both these conclusions are used later in the design of the exhaust port or valve areas.

Notice from (7.20) that the mass discharged is related to the volume change as well as the pressure change. The smaller the volume change, the smaller the final mass m_2, whence we should design for as rapid a port opening as possible.

In a four-stroke engine V_2 will correspond to the volume at the end of the stroke. In a two-stroke engine the volume V_2 will correspond to the cylinder volume at inlet port opening.

7.3.2 Exhaust Stroke

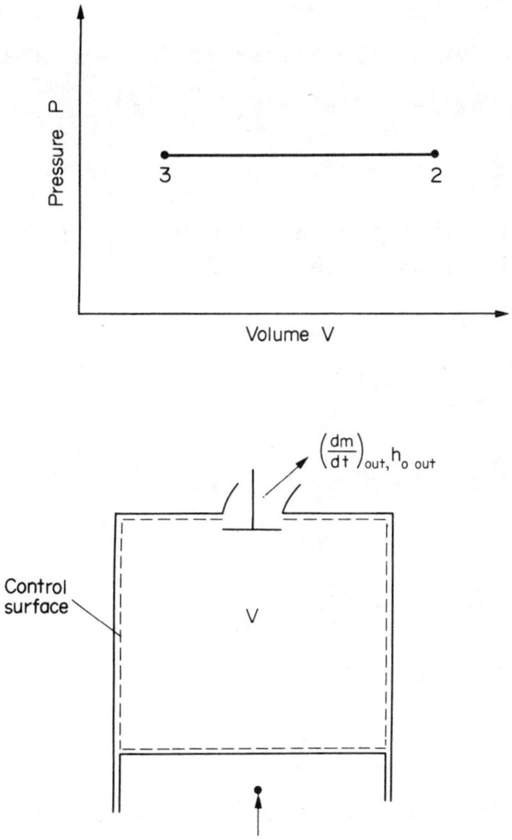

FIG. 7.9. Exhaust stroke. m, mass in cylinder; V, volume; e, specific internal energy; h, specific enthalpy

In Fig. 7.9 the pressure/volume diagram during the exhaust stroke is shown. This process only occurs in the four-stroke engine. At the beginning of the stroke the cylinder volume is V_2 and at the end V_3. The cylinder pressure p is assumed to be constant.

Subscript 2 refers to conditions at the beginning of the stroke and subscript 3 at the end. We shall assume an adiabatic process, as before, that is, $dQ/dt = 0$. There is no shaft work in the control system, whence $dW_s/dt = 0$. No gas flows into the cylinder and $(dm/dt)_{in} = 0$. Hence the first law (2.32) for the control system is

$$- p \frac{dV}{dt} = \frac{\partial(me)}{\partial t} + \left(\frac{dm}{dt}\right)_{out} h_{o\ out} \qquad (7.21)$$

and the continuity equation is

$$\left(\frac{dm}{dt}\right)_{cv} = - \left(\frac{dm}{dt}\right)_{out}. \qquad (7.22)$$

For a time step dt during the stroke it follows that

$$- p\ dV = d(me) + h_{o\ out}\ dm_{out}, \qquad (7.23)$$

$$dm_{out} = - dm. \qquad (7.24)$$

Once again we assume that the enthalpy $h_{o\ out}$ of the gas stream leaving the cylinder equals the cylinder enthalpy h, that is,

$$h = h_{o\ out}.$$

If we substitute for $h_{o\ out}$, dm_{out} into the first law (7.23) and expand, we obtain as before

$$- \frac{p\ dV}{m(h-e)} = \frac{de}{(h-e)} - \frac{dm}{m}, \qquad (7.25)$$

which reduces to

$$- \frac{dV}{V} = \frac{dT}{(\kappa-1)T} - \frac{dm}{m}. \qquad (7.26)$$

This we can rearrange to give

$$\frac{dm}{m} - \frac{dV}{V} = \frac{dT}{(\kappa-1)T}. \qquad (7.27)$$

Now the cylinder pressure p is constant and the state equation is

$$pV = mRT$$

for the process, then

$$\frac{dV}{V} = \frac{dm}{m} + \frac{dT}{T} \qquad (7.28)$$

since $dp = 0.$

Substituting (7.28) into (7.27) we have

$$- \frac{dT}{T} = \frac{dT}{(\kappa-1)T}. \qquad (7.29)$$

This identity can only be satisfied if dT = 0. That is the process is a <u>constant enthalpy</u> process.

An alternative proof is by the definition of dh:

$$dh = v\,dp.$$

Now $$dH = m\,dh = mv\,dp,$$

where v is the specific volume.

Hence $$dH = V\,dp.$$

Since the volume V is finite when dp = 0, dH = 0 and, since m is not zero, it follows dh = 0.

For an adiabatic constant pressure exhaust stroke the cylinder temperature is constant.

7.3.3 Suction Stroke

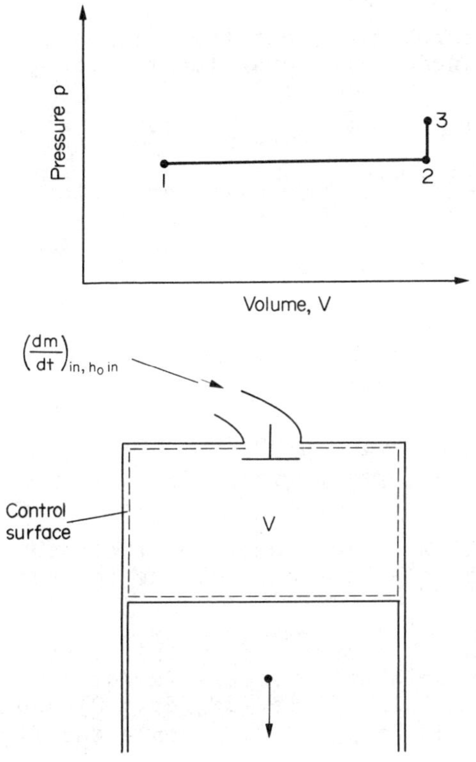

FIG. 7.10. Suction stroke. m, mass in cylinder; V, volume.

In Fig. 7.10 the pressure/volume diagram during the suction stroke is shown. This process only occurs in a four-stroke engine. At the beginning of the stroke the cylinder volume is V_1. The piston moves outwards and the cylinder volume increases up to V_2 with air entering the cylinder. The air pressure will be higher than the cylinder pressure in order that flow can take place across the inlet valve. At the beginning of the suction stroke the clearance volume V_1 may or may not contain the residuals from the previous exhaust stroke. This will depend on the valve timing and flow areas. If the inlet and exhaust valves are open at the same time, the inlet air may pass straight through to the exhaust, thus scavenging the clearance volume. We shall consider the general case with the possibility of residuals. In this case the cylinder composition will vary during the suction stroke. The analysis is therefore slightly different from the exhaust stroke.

Let the internal energy of the gases in the control volume be E_{cv}, where $E_{cv} = \Sigma(me)$ and m is the mass of a species whose specific internal energy is e. We shall now assume the specific heats of all the constituents are the same and that they are constant. Then

$$me = m\,C_v T \quad \text{and} \quad E_{cv} = C_v \Sigma\, mT.$$

The pressure is uniform throughout the cylinder. If v is the specific volume of one of the constituents, the general gas flow will be

$$\Sigma\, mT = \frac{\Sigma\, mpv}{R} = \frac{p}{R}\,\Sigma\, mv = \frac{pV}{R}.$$

Hence the internal energy of the cylinder contents will be

$$E_{cv} = \frac{C_v}{R}\, pV.$$

Now

$$\frac{C_v}{R} = \frac{1}{\kappa - 1}$$

whence

$$\frac{\partial E_{cv}}{\partial t} = \frac{1}{\kappa-1}\,\frac{\partial (pV)}{\partial t}. \tag{7.30}$$

Expression (7.30) allow us to determine the internal energy change in the cylinder <u>without</u> a knowledge of the composition of the cylinder contents.

We are now in a position to apply the first law (2.32). As before, we assume an adiabatic process ($dQ/dt = 0$). There is no shaft work at the control surface ($dW_s/dt = 0$) and no gas flows out of the cylinder and $(dm/dt)_{out} = 0$. Hence the first law becomes

$$-\,p\,\frac{dV}{dt} = \frac{1}{\kappa-1}\,\frac{\partial (pV)}{\partial t} - \left(\frac{dm}{dt}\right)_{in} h_{o\ in}. \tag{7.31}$$

We can expand (7.31) to obtain

$$-p\frac{dV}{dt} = \frac{1}{\kappa-1}\left(p\frac{dV}{dt} + V\frac{dp}{dt}\right) - \left(\frac{dm}{dt}\right)_{in} h_{o\ in} \qquad (7.32)$$

or rearrange

$$\left(\frac{dm}{dt}\right)_{in} h_{o\ in} - \frac{V}{\kappa-1}\frac{dp}{dt} = \frac{\kappa}{\kappa-1}\,p\,\frac{dV}{dt}\,. \qquad (7.33)$$

The continuity equation is

$$\left(\frac{dm}{dt}\right)_{cv} = \left(\frac{dm}{dt}\right)_{in}. \qquad (7.34)$$

The suction stroke may be considered to take place in two phases: a constant pressure phase from V_1 to V_2, and a constant volume phase from p_2 to p_3.

We shall first consider the step 1 to 2.

In this step $dp/dt = 0$ and it follows from (7.33) that

$$\left(\frac{dm}{dt}\right)_{in} h_{o\ in} = \frac{\kappa}{\kappa-1}\,p\,\frac{dV}{dt}\,. \qquad (7.35)$$

Now $$h_{o\ in} = C_p\,T_{o\ in}$$

and $$\frac{\kappa}{\kappa-1} = \frac{C_p}{R},$$

whence (7.35) becomes

$$C_p\,T_{o\ in}\frac{dm_{in}}{dt} = \frac{C_p}{R}\,p\,\frac{dV}{dt} \qquad (7.36)$$

or $$T_{o\ in}\frac{dm_{in}}{dt} = \frac{p}{R}\frac{dV}{dt}\,. \qquad (7.37)$$

For a time step dt,

$$T_{o\ in}\,dm_{in} = \frac{p}{R}\,dV. \qquad (7.38)$$

To obtain the relationships between the conditions at the beginning (1), and the end (2) of the process 1→2 we integrate (7.38) noting that

$$m_{in} = \int dm_{in} \quad \text{and} \quad p = \text{constant.}$$

Then $$T_{o\ in}\,m_{in} = \frac{p}{R}\,(V_2 - V_1)\,. \qquad (7.39)$$

Now $$pV = mRT$$

and the mass at 2 is the sum of the inflow mass m_{in} and the mass at 1; thus

$$m_2 = m_1 + m_{in}.$$

Hence expression (7.39) becomes

$$T_{o\ in}\, m_{in} = \frac{1}{R}\left(m_2 R T_2 - m_1 R T_1\right)$$

or

$$T_{o\ in}\, m_{in} = m_2 T_2 - m_1 T_1$$

or

$$T_{o\ in}\, m_{in} = \left(m_1 + m_{in}\right) T_2 - m_1 T_1$$

or

$$T_2 = \frac{T_{o\ in}\, m_{in} + T_1 m_1}{\left(m_1 + m_{in}\right)}. \qquad (7.40)$$

This gives the temperature T_2 at the end of the period $1 \rightarrow 2$.

We can rearrange (7.40) to give

$$T_2 = T_{o\ in} + \left[\frac{m_1}{m + m_{in}}\right]\left(T_1 - T_{o\ in}\right). \qquad (7.41)$$

Let us examine this expression. If the clearance volume had been scavenged before the suction stroke commenced such that $T_1 = T_{o\ in}$, then the temperature at the end of the phase $1 \rightarrow 2$ would be equal to T_1 or $T_{o\ in}$. Hence the suction process $1 \rightarrow 2$ would be a constant temperature process. If, on the other hand, the temperature T_1 in the clearance volume is greater than $T_{o\ in}$, then the final temperature T_2 would be greater than $T_{o\ in}$. The magnitude of the final temperature would depend on the mass of air entering the cylinder m_{in} and the mass of the residuals m_1.

We will now consider the second phase $2 \rightarrow 3$. In this phase the volume is constant and

$$\frac{dV}{dt} = 0.$$

The expression (7.33) now becomes

$$\left(\frac{dm}{dt}\right)_{in} h_{o\ in} - \frac{V}{\kappa - 1}\frac{dp}{dt} = 0$$

or

$$\frac{V}{\kappa - 1}\frac{dp}{dt} = C_p\, T_{o\ in}\left(\frac{dm}{dt}\right)_{in}. \qquad (7.42)$$

For a time step dt,

$$\frac{V}{\kappa - 1}\, dp = C_p\, T_{o\ in}\, dm_{in}. \qquad (7.43)$$

To obtain the relationship between the conditions at 2 and 3 we integrate (7.43) and obtain

$$\frac{V}{\kappa-1}(p_3-p_2) = C_p T_{o\ in} m_{in}. \tag{7.44}$$

Now

$$p_3V_3 = m_3RT_3,$$

$$p_2V_2 = m_2RT_2,$$

$$V = V_2 = V_3,$$

$$m_{in} = m_3-m_2,$$

and

$$C_p = \frac{\kappa}{\kappa-1} R.$$

Hence substituting into (7.44) we obtain, after simplification,

$$T_3 = T_2 + \frac{m_{in}}{m_{in}+m_2}\left(\kappa T_{o\ in}-T_2\right). \tag{7.45}$$

We shall now examine the special case of an engine with valve overlap with 100% scavenged air. Under these conditions the temperature in the cylinder at the beginning of the suction stroke T_1 will be equal to the air inlet temperature $T_{o\ in}$.

We have shown that under these conditions $T_2 = T_1$. Hence T_2 equals the air inlet temperature $T_{o\ in}$. If we substitute into (7.45), we have

$$T_3 = T_{o\ in} + \frac{m_{in}}{m_{in}+m}\left(\kappa T_{o\ in}-T_{o\ in}\right)$$

or

$$T_3 = T_{o\ in}\left[1 + \frac{m_{in}}{m_{in}+m_2}(\kappa-1)\right]. \tag{7.46}$$

If we let the air in m_{in} from 1 to 2 be equal to $(m_{in})_{12}$, and for 2 to 3 be equal to $(m_{in})_{23}$, it follows that the total air drawn into the engine in the suction stroke will be

$$m_{tot} = \left(m_{in}\right)_{12} + \left(m_{in}\right)_{23}.$$

Now

$$m_2 = m_1 + \left(m_{in}\right)_{12}$$

and

$$\left(m_{in}\right)_{23}+m_2 = m_1 + \left(m_{in}\right)_{12} + \left(m_{in}\right)_{23} = m_1 + m_{tot}. \tag{7.47}$$

If we note that in (7.46) $m_{in} = \left(m_{in}\right)_{23}$, it follows that substituting (7.47) into (7.46) we have

$$T_3 = T_{o\ in}\left[1 + \frac{(m_{in})_{23}\ (\kappa-1)}{(m_1+m_{tot})}\right]. \qquad (7.48)$$

Now the term in the brackets is always greater than unity. Thus the trapped temperature T is greater than the inlet air with 100% scavenged clearance volume. The increase in the cylinder temperature above the inlet air temperature is due to the transformation of the flow work of the air entering the cylinder into internal energy during the constant volume phase 2⟶3.

In the analysis given above we have assumed an adiabatic process. In practice there will be heat transfer taking place from the cylinder wall to the air entering the cylinder. This will have a significant effect on the volumetric efficiency, particularly in a naturally aspirated engine. Furthermore, in a naturally aspirated engine the scavenge effects due to the overlap are minimal. Thus the trapped charge temperature will be influenced both by the heat transfer and the residual temperature. We shall examine these effects separately.

We shall first examine the heat transfer effects. For this purpose we shall assume 100% scavenge of the clearance volume and the temperature is then uniform throughout the cylinder. In a naturally aspirated engine the volumetric efficiency η_V is based on the swept volume V_S. For 100% scavenge efficiency the volumetric efficiency equals the charging efficiency η_{CH}.

We shall consider the process to follow from 1⟶2 in Fig. 7.10. The first law for the control volume will be

$$\frac{dQ}{dt} - p\frac{dV}{dt} = \frac{\partial E_{cv}}{\partial t} - \left(\frac{dm}{dt}\right)_{in} h_{o\ in} \qquad (7.49)$$

since dW_s/dt and $(dm/dt)_{out}$ are zero.

Since the temperature in the cylinder is uniform

$$\frac{\partial E_{cv}}{\partial t} = \frac{\partial(me)}{\partial t}$$

where m is the mass in the cylinder of specific internal energy e. In a time interval dt the first law becomes

$$dQ - p\ dV = d(me) - dm_{in}\ h_{o\ in}. \qquad (7.50)$$

For the suction stroke 1⟶2,

$$Q - p(V_2 - V_1) = m_2e_2 - m_1e_1 - m_{in}\ h_{o\ in}, \qquad (7.51)$$

where Q is the heat transfer during the stroke, p is the constant cylinder pressure and m_{in} the quantity of air supplied during the stroke.

Now $$pV_2 = p_2V_2 = m_2RT_2$$

and $$pV_1 = p_1V_1 = m_1RT_1.$$

Hence $$Q-R\left(m_2T_2 - m_1T_1\right) = m_2e_2 - m_1e_1 - m_{in}\, h_{o\ in}.$$

Rearranged, we obtain

$$Q + m_{in}\, h_{o\ in} = m_2\left(e_2+RT_2\right) - m_1\left(e_1+RT_1\right),$$

$$Q + m_{in}\, h_{o\ in} = m_2h_2 - m_1h_1 = C_p\left(m_2T_2 - m_1T_1\right) \quad (7.52)$$

or $$\frac{Q}{C_p} + m_{in}\, T_{o\ in} = m_2T_2 - m_1T_1 = \frac{p}{R}\,(V_2-V_1). \quad (7.53)$$

Now the swept volume $V_s = V_2-V_1$.

Hence dividing (7.53) by $T_{o\ in}$ and substituting for V_2-V_1 we obtain, after rearrangement,

$$m_{in} = \frac{p\ V_s}{R\ T_{o\ in}} - \frac{Q}{C_p\ T_{o\ in}}\ . \quad (7.54)$$

The <u>maximum</u> amount of air which could be drawn into the cylinder at the <u>cylinder</u> pressure p during the suction stroke would be

$$m_a = \frac{p\ V_s}{R\ T_{o\ in}}. \quad (7.55)$$

Let m_q be the loss in trapped mass due to heat transfer;

then $$m_q = \frac{Q}{C_p\ T_{o\ in}} \quad (7.56)$$

and $$m_{in} = m_a - m_q.$$

The charging efficiency based on the swept volume V_s is

$$\eta_{CH} = \frac{m_{in}}{\left(m_{in}\right)_{max}}\ . \quad (7.57)$$

$\left(m_{in}\right)_{max}$ is the <u>maximum</u> quantity of air which could be drawn in the cylinder at <u>inlet pressure</u> p_{in}, that is

$$\left(m_{in}\right)_{max} = \frac{p_{in}\ V_s}{R\ T_{o\ in}} \quad (7.58)$$

If we define η'_{CH} as the charging efficiency <u>without</u> heat transfer, then

$$\eta'_{CH} = \frac{m_a}{(m_{in})_{max}}, \tag{7.59}$$

and if we let Δq be the ratio,

$$\Delta q = \frac{m_q}{(m_{in})_{max}}$$

$$= \frac{Q}{C_p T_{o\ in}} \frac{RT_{o\ in}}{p_{in} V_s}$$

$$= \frac{\kappa-1}{\kappa} \frac{Q}{V_s} \frac{1}{p_{in}},$$

$$\Delta q = \frac{\kappa-1}{\kappa} \frac{q}{p_{in}}, \tag{7.60}$$

where q is the heat transfer per unit volume.

Then the charging efficiency η_{CH} is given by

$$\eta_{CH} = \frac{m_a}{(m_{in})_{max}} - \frac{m_q}{(m_{in})_{max}}$$

or

$$\eta_{CH} = \eta'_{CH} - \frac{\kappa-1}{\kappa} \frac{q}{p_{in}}. \tag{7.61}$$

The parameter q is the heat transferred to the incoming air per unit swept volume. Thus the charging efficiency decreases linearly with the heat transferred rate per unit volume. For 100% scavenging of the clearance volume the charging efficiency equals the volumetric efficiency. It follows therefore that the greater the heat addition the lower the volumetric efficiency.

An approximate assessment of the effect of the temperature rise (ΔT) of the incoming air on the volumetric efficiency may be made by a simple analysis. We assume that all the heat addition is made to the incoming air $(m_{in} = \rho_{in} V_s)$. Then

$$Q \simeq m_{in} C_p \Delta T \simeq \rho_{in} V_s C_p \Delta T.$$

Now

$$p_{in} = R\rho_{in} T_{o\ in},$$

whence

$$\frac{\kappa-1}{\kappa} \frac{q}{p_{in}} = \frac{\kappa-1}{\kappa} \frac{\rho_{in} V_s C_p \Delta T}{V_s} \frac{1}{R\rho_{in} T_{o\ in}} = \frac{\Delta T}{T_{o\ in}}$$

The volumetric efficiency is then

$$\eta_V = \eta_V' - \frac{\Delta T}{T_{o\ in}}. \tag{7.62}$$

The percentage drop in volumetric efficiency is approximately equal to the percentage rise in air temperature in the suction stroke. This approximation is reasonable up to about 15% of the changes, and gives a feel for the influence of air temperature rise on volumetric efficiency.

Unless there is valve overlap it is not possible to scavenge the products of combustion from the clearance volume. The final temperature of the trapped charge will be influenced by the volume of the clearance at top dead centre and the temperature of the residuals in the clearance volume. A fairly simple analysis may be made using the first law to examine this effect. We shall neglect heat transfer and assume a constant cylinder pressure during the suction process. In this case from (7.52) we have

$$m_{in}\, h_{o\ in} = m_2 h_2 - m_1 h_1 \tag{7.63}$$

since $Q = 0.$

Now $m_2 = m_{in} + m_1$

and $h = C_p T.$

It follows therefore that by rearranging (7.63) we have

$$T = \frac{T_{o\ in} + \dfrac{m_1}{m_{in}} T_1}{\left(1 + \dfrac{m_1}{m_{in}}\right)} \tag{7.64}$$

Now
$$\frac{m_1}{m_{in}} = \frac{p_1 V_1}{T_1} \frac{T_{o\ in}}{p_{in} V_s}. \tag{7.65}$$

The compression ratio r is given by

$$r = \frac{V_1 + V_s}{V_1} \quad \text{and} \quad \frac{V_s}{V_1} = r-1,$$

whence
$$\frac{m_1}{m_{in}} = \left(\frac{p_1}{p_{in}}\right)\left(\frac{T_{o\ in}}{T_1}\right)\left(\frac{1}{r-1}\right)$$

and

$$T_2 = \frac{T_{o\ in}\left(1 + \left(\frac{p_1}{p_{in}}\right)\left(\frac{1}{r-1}\right)\right)}{1 + \frac{p_1}{p_{in}}\left(\frac{1}{r-1}\right)\left(\frac{T_{o\ in}}{T_1}\right)}$$

or

$$T_2 = T_{o\ in}\left[\frac{(r-1) + \frac{p_1}{p_{in}}}{(r-1) + \left(\frac{p_1}{p_{in}}\right)\left(\frac{T_{o\ in}}{T_1}\right)}\right] \tag{7.66}$$

If there is no pressure drop across the ports, then

$$T_2 = T_{o\ in}\left[\frac{r}{(r-1) + \frac{T_{o\ in}}{T_1}}\right] \tag{7.67}$$

In Fig. 7.11 the ratio of the trapped to inlet air temperatures is shown for various residual temperature to inlet temperature ratios. It will be seen that the trapped temperature increases with increase in residual temperatures and decreases in compression ratio. To a first order

$$\frac{d\,\eta_V}{\eta_V} \doteqdot \frac{dp}{p_{in}} - \frac{dT}{T_{in}} \,. \tag{7.68}$$

Hence the effect of the residual temperature on volumetric systems is more significant on low compression ratio engines (gasolene spark ignition engines) than high compression ratio engines (diesel compression ignition engines).

7.4 SCAVENGE PROCESS

During the scavenge period the cylinder is exposed to both the inlet and exhaust systems at the same time. The flow processes are extremely complex and we have to make simplifying assumptions. A number of models have been developed to describe the scavenge process, each of greater complexity. Basically we consider (i) isothermal models in which the process is at constant pressure and temperature, and (ii) gas dynamics models in which allowance is made for the variation in temperature and pressure. Isothermal models are simple and lead to quantitative assessment of the scavenge process which can be tested experimentally in a flow visualization rig using photography and sampling techniques. Some of these rigs use dyed water to represent the products of combustion and the fresh charge, others use air and carbon dioxide or other dissimilar gases. Tracer methods are also used. We shall first discuss the isothermal models and then examine one simple gas dynamic model. In

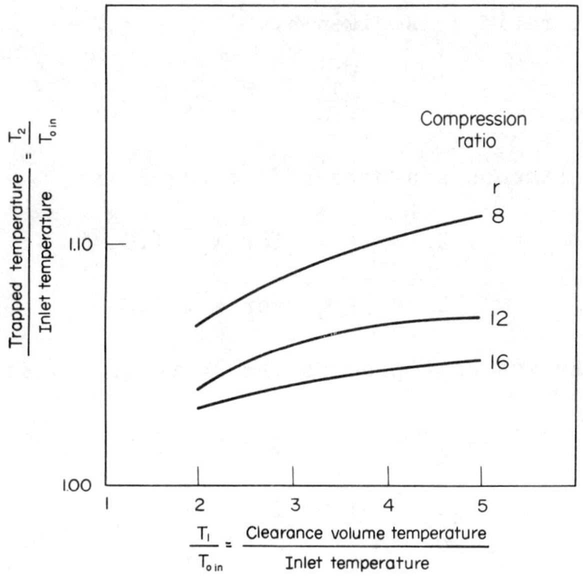

FIG. 7.11. Influence of clearance volume temperature and compression ratio on trapped temperature (neglecting pressure drop across inlet valve).

the isothermal models the mass flow rates are proportioned to the volume flow rates and the efficiencies are normally expressed in volume terms.

7.4.1 Isothermal Scavenge Models

A large mass of the products of combustion leaves the cylinder during the exhaust blowdown period, the remainder is evacuated during the scavenge process. The simplest model of the scavenge process is displacement scavenging. As the term implies, the incoming charge of fresh air (plus fuel in a spark ignition engine) displaces the products of combustion. There is no mixing in the cylinder. The efficiency of the scavenge process is directly related to the air supplied.

Let Z be the instantaneous scavenge efficiency given by the ratio of the volume of air in the cylinder $V_{a'}$ to the cylinder volume V. Then the instantaneous efficiency is

$$Z = \frac{V_{a'}}{V} \tag{7.69}$$

Since the incoming air displaces the cylinder products, the air supplied at any instant of time V_a will equal $V_{a'}$ until V_a exceeds the cylinder volume V when air will pass into the exhaust pipe. Hence

$$V_{a'} = V_a. \tag{7.70}$$

Now the scavenge ratio λ is given by

$$\lambda = \frac{V_a}{V} \tag{7.71}$$

Hence the instantaneous scavenge efficiency Z is:

$$Z = \lambda \quad \text{for } \lambda \leq 1.0, \tag{7.72a}$$

$$Z = 1 \quad \text{for } \lambda > 1.0. \tag{7.72b}$$

At the end of the scavenge process the scavenge efficiency η_{SC} will equal Z. Hence

$$\eta_{SC} = \lambda \quad \text{for } \lambda < 1.0 \tag{7.73a}$$

or

$$\eta_{SC} = 1.0 \quad \text{for } \lambda > 1.0. \tag{7.73b}$$

For displacement scavenging, the scavenge efficiency is equal to scavenge ratio up to a scavenge ratio of 1.0 whence it remains at 1.0 for scavenge ratios exceeding unity (Fig. 7.12).

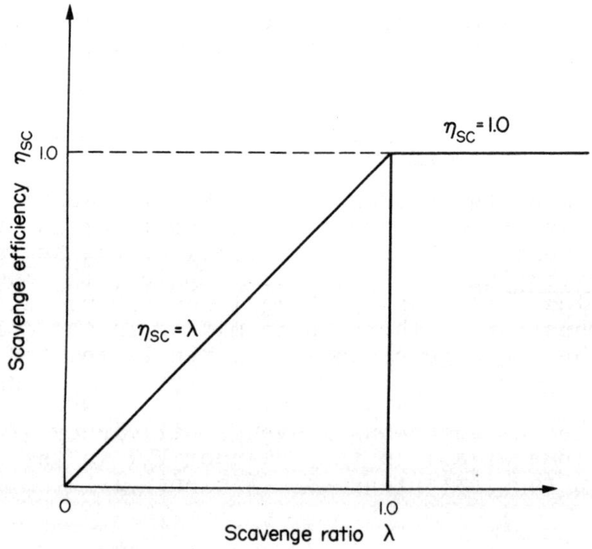

FIG. 7.12. Displacement scavenging—scavenge efficiency.

For an isothermal process the charging efficiency η_{CH} equals the scavenge efficiency, whence

$$\eta_{CH} = \eta_{SC} = \lambda \quad \text{for } \lambda < 1.0 \tag{7.74a}$$

and

$$\eta_{CH} = \eta_{SC} = 1.0 \quad \text{for } \lambda > 1.0. \tag{7.74b}$$

Displacement scavenging model is the upper bound for the scavenge process. In practice the scavenge process is not as efficient as displacement scavenging. Another model suggested by Hopkinson[2] is called mixing scavenging (Fig. 7.13). The following assumptions are made in this model:

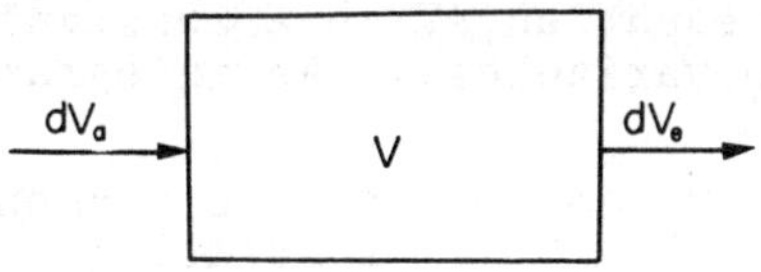

FIG. 7.13. Mixing scavenging.

(i) The process occurs at a constant cylinder volume V.

(ii) The densities of the air and exhaust gas are the same (i.e. it is a constant pressure, isothermal process).

(iii) The addition of a volume of air dV_a to the cylinder causes an equal volume of exhaust gases dV_e to leave the cylinder.

(iv) The process is one of perfect mixing. The air enters the cylinder contents, and a mixture of gas and air leave the cylinder at the same time.

(v) The instantaneous scavenge efficiency is given by $Z = V_{a'}/V$, where $V_{a'}$ is the volume of air retained in the cylinder. Hence Z is the proportion of air in the cylinder contents at any instant of time.

The volume of air entering the cylinder at any instant of time is dV_a, the volume of air plus gas leaving is dV_e; of this the volume of air leaving the cylinder at any instant of time is ZdV_e. The volume of air retained in the cylinder is $dV_{a'}$.

Thus
$$dV_{a'} = dV_a - ZdV_e \tag{7.75}$$

and
$$Z = \frac{V_{a'}}{V}. \tag{7.76}$$

Hence
$$V\,dZ = dV_{a'} \tag{7.77}$$

Substituting (7.77) into (7.75),

$$V\,dz = dV_a - Z\,dV_e. \tag{7.78}$$

Now the <u>volume</u> of air entering the cylinder dV_a equals the volume of gas plus air leaving the cylinder dV_e. Substituting for dV_e in (7.78) we have

$$V\,dZ = dV_a - Z\,dV_a = dV_a(1-Z)$$

or

$$\frac{dZ}{1-Z} = \frac{dV_a}{V} \tag{7.79}$$

In this differential equation, V is a constant (the cylinder volume), Z and V_a are variables. We integrate between the limits:

$Z = 0,\ V_a = 0$ at the commencement of scavenging;

$Z = \eta_{SC},\ V_a = V_a$ at the end of scavenging where V_a is now the <u>total</u> air supplied.

Using the scavenge ratio λ for a constant pressure and temperature process,

$$\lambda = \frac{V_a}{V}.$$

The second limit is

$$Z = \eta_{SC}, \quad V_a = \lambda V.$$

Integration of (7.79) is then

$$\int_o^{\eta_{SC}} \frac{dZ}{1-Z} = \int_o^{\lambda V} \frac{dV_a}{V} \tag{7.80}$$

$$-\ell n(1-\eta_{SC}) = \lambda$$

and

$$\eta_{SC} = 1 - e^{-\lambda}. \tag{7.81}$$

For an isothermal process at constant pressure,

$$\eta_{CH} = \eta_{SC}$$

and

$$\eta_{CH} = 1 - e^{-\lambda}. \tag{7.82}$$

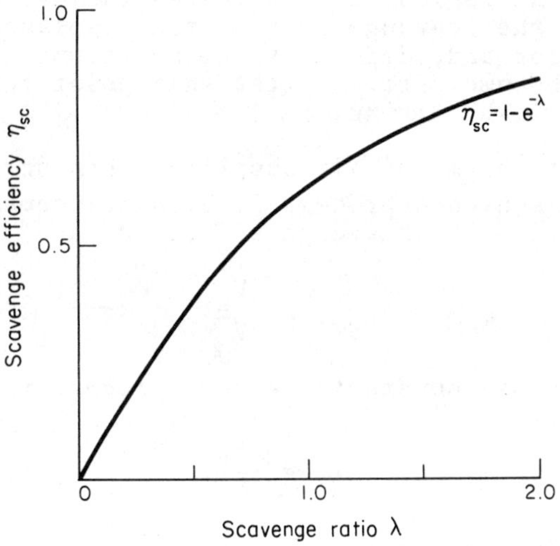

FIG. 7.14 Mixing scavenging—scavenge efficiency.

This model produces the curve shown in Fig. 7.14 for the variation of scavenge or charging efficiency with scavenge ratio. You will observe that at low scavenge ratios there is a rapid increase of efficiency with λ, but after about a value of λ equal to unity the increase in efficiency falls off. Physically this is an unrealistic model since there is inadequate time for the mixing process to take place simultaneously in the cylinder, nor is it aerodynamically possible since the mixing can only occur at the jet boundaries of the incoming charge. Benson and Brandham[3] suggested a model combining mixing and displacement. This is called the mixing-displacement scavenging model. In this model we subdivide the cylinder into two regions: a mixing region and a displacement region (Fig. 7.15).

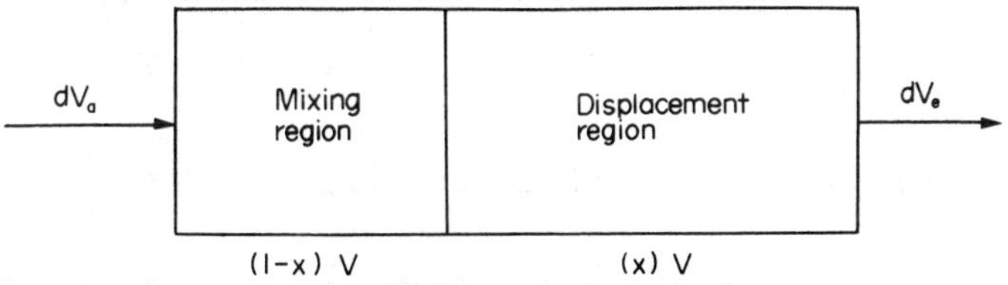

Fig. 7.15. Mixing-displacement scavenge model.

We assume that air enters the cylinder and mixes with the cylinder contents in the mixing region at the same time gas leaves the cylinder from the displacement region. As for the first model, we assume a constant pressure constant temperature process. If the cylinder volume is V and x the fraction of the cylinder volume occupied initially by the displacement region, the initial volumes of the two regions are:

displacement region	xV,
mixing region	$(1-x)V$.

In the displacement region the gas leaves the cylinder without mixing. During the scavenge process the displacement region becomes smaller and smaller and, dependent on the volume of air supplied, will eventually become zero. After this point the whole of the scavenge process is by mixing.

If V_a is the volume of air supplied, then until V_a exceeds xV the whole of the scavenge process is by displacement. Thus for $V_a \leq xV$ we can write the scavenge efficiency as

$$\eta_{SC} = \frac{V_a}{V}$$

Now for an isothermal constant pressure process the scavenge ratio is

$$\lambda = \frac{V_a}{V}.$$

Hence we can write for $\lambda \leq x$

$$\lambda_{SC} = \lambda. \tag{7.83}$$

When the air supply V_a exceeds the displacement volume xV, the volume of the mixing region is V and we can compute the scavenge efficiency by the Hopkinson method using (7.79):

$$\frac{dZ}{1-Z} = \frac{dV_a}{V}.$$

The limits of integration are now:

(i) when the displacement scavenge is completed (i.e. mixing commences for the whole cylinder),

$$Z = x, \quad V_a = xV;$$

(ii) at the end of the whole scavenge process,

$$Z = \eta_{SC}, \quad V_a = \lambda V.$$

Integration of (7.79) is then:

$$\int_x^{\eta_{SC}} \frac{dZ}{1-Z} = \int_{xV}^{\lambda V} \frac{dV_a}{V}, \tag{7.84}$$

$$-\ell n\left(\frac{1-\eta_{SC}}{1-x}\right) = \lambda - x$$

or $$\eta_{SC} = 1-(1-x)e^{(x-\lambda)} \tag{7.85}$$

We can define two scavenge efficiencies depending on the scavenge ratio λ:

when $\lambda \leq x$ $$\eta_{SC} = \lambda \tag{7.86a}$$

and when $\lambda > x$ $$\eta_{SC} = 1-(1-x)e^{(x-\lambda)}. \tag{7.86b}$$

As before, for constant pressure isothermal process

$$\eta_{CH} = \eta_{SC}.$$

The resultant curves are shown in Fig. 7.16.

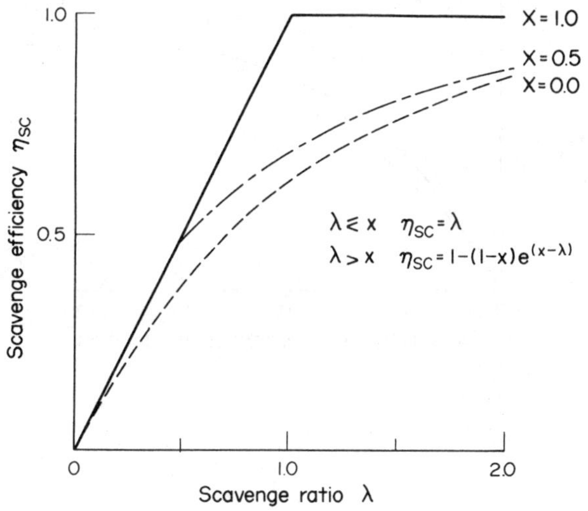

FIG. 7.16. Mixing-displacement scavenging—scavenge efficiency.

The Benson-Brandham model gives a scavenge efficiency between displacement and mixing scavenging. In practice there will be some short circuiting of the fresh charge direct to the exhaust pipe. We can allow for this by a simple modification. If we assume that the short circuiting air is a direct proportion of the scavenge air, say $y\lambda$. Then we replace λ by $(1-y)\lambda$ in expressions (7.86a and b), and we have two scavenge efficiencies as before:

$(1-y)\lambda \leq x$ $$\eta_{SC} = (1-y)\lambda, \tag{7.87a}$$

$(1-y)\lambda > x$ $$\eta_{SC} = 1-(1-x)e^{(x-(1-y)\lambda)}. \tag{7.87b}$$

The scavenging efficiency for the three models with short-circuiting is shown in Fig. 7.17.

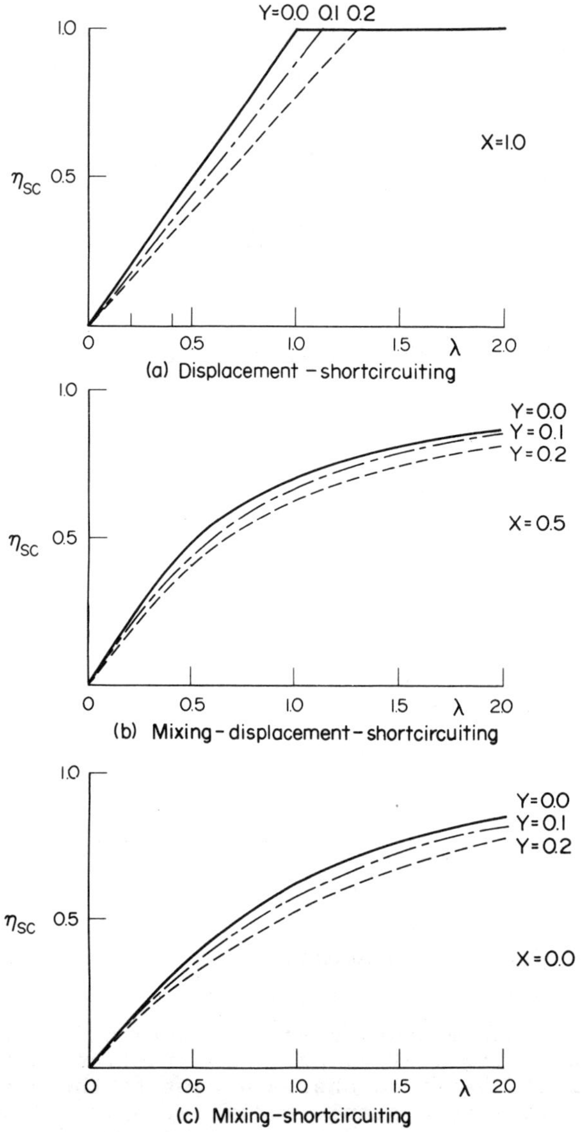

FIG. 7.17. Mixing-displacement-shortcircuiting scavenging.

All four scavenge models can be represented by the following expressions and conditions:

when $(1-y)\lambda \leq x$ $\quad \eta_{sc} = (1-y)\lambda,$ (7.88a)

when $(1-y)\lambda > x$ $\quad \eta_{sc} = 1-(1-x)e^{(x-(1-y)\lambda)},$ (7.88b)

$\eta_{sc} > 1$ $\quad \eta_{sc} = 1.0.$ (7.88c)

The pure mixing model corresponds to $x = 0.0$, $y = 0.0$. The pure displacement model corresponds to $x = 1.0$, $y = 0.0$. The mixing-displacement model corresponds to

$$0.0 < x < 1.0, \; y = 0.$$

The mixing-displacement-short circuit model corresponds to

$$0.0 < x < 1.0, \; 0.0 < y < 1.0.$$

7.4.2 Non-isothermal Scavenge Models

In the non-isothermal models we apply the general first law to the cylinder control volume (2.32). The first law is

$$\frac{dQ}{dt} - \frac{dW_s}{dt} - p\frac{dV}{dt} = \left(\frac{\partial E}{\partial t}\right)_{cv} + \left(\frac{dm}{dt}\right)_{out} h_{o\ out} - \left(\frac{dm}{dt}\right)_{in} h_{o\ in}. \qquad (2.32)$$

For ease of analysis we use the following symbols:

Air entering the cylinder

$$\left(\frac{dm}{dt}\right)_{in} = \dot{m}_a, \qquad h_{o\ in} = h_a.$$

Exhaust products leaving the cylinder:

$$\left(\frac{dm}{dt}\right)_{out} = \dot{m}_e, \qquad h_{o\ out} = h_e.$$

We shall assume that the cylinder volume V is constant and the pressure p is also constant. We shall further assume that the air entering the cylinder mixes with the cylinder contents and a mixture of air and gas of the same composition as the cylinder. Thus we shall use a mixing model which is non-isothermal.

We shall make the following thermodynamic assumptions:

(i) ideal gas, $pV = mRT$, with constant specific heats;

(ii) $h = e = 0$ at $T = 0$

whence $h = \frac{\kappa}{\kappa-1} RT, \qquad e = \frac{RT}{\kappa-1};$

(iii) the process is adiabatic.

The internal energy of the contents E are given by

$$E = me = \frac{mRT}{\kappa-1} = \frac{pV}{\kappa-1}.$$

For the control volume therefore the rate of change of internal energy $(\partial E/\partial t)_{cv}$ will be given by

$$\left(\frac{\partial E}{\partial t}\right)_{cv} = \frac{\partial}{\partial t}\left(\frac{pV}{\kappa-1}\right).$$

Now since p and V are both constant it follows that

$$\left(\frac{\partial E}{\partial t}\right)_{cv} = 0.$$

For the control volume $dW_s/dt = 0$ and, since the process is adiabatic, $dQ/dt = 0$.

The first law for the control volume (2.32) is then

$$0 = \dot{m}_e h_e - \dot{m}_a h_a.$$

If we consider a time element dt we have

$$0 = dm_e h_e - dm_a h_a,$$

$$dm_a h_a = dm_e h_e. \qquad (7.89)$$

The <u>mass</u> balance for the volume is

$$dm = dm_a - dm_e$$ (dm = change of mass <u>in</u> the cylinder)

and $$h = h_e.$$

Substitute for dm_e and h_e in (7.89) we have

$$dm_a h_a = h(dm_a - dm)$$

or $$dm_a(h - h_a) = h\ dm. \qquad (7.90)$$

Now $$me = \frac{mRT}{\kappa-1} = \frac{pV}{\kappa-1},$$

$$mh = \frac{\kappa}{\kappa-1}mRT = \frac{\kappa}{\kappa-1}\ pV.$$

Since p, V are constant,

$$d(mh) = 0$$

or $$h\ dm = -\ m\ dh.$$

Substitute for h dm into (7.90) we have

$$dm_a(h-h_a) = -m\ dh$$

or

$$dm_a = -\frac{m\ dh}{(h-h_a)}. \qquad (7.91)$$

If we multiply the numerator and denominator of the right-hand side of the above expression by h we have

$$dm_a = -\frac{mh\ dh}{h(h-h_a)} \qquad (7.92)$$

We can integrate the above expression between inlet air port opening (APO_1) and inlet air port closing (APC_2). The suffices 1, 2 will correspond to the states at APO, APC, respectively. We note that

$$\int_1^2 dm_a = m_a = \text{the mass of air supplied}$$

$$mh = \text{constant.}$$

Thus the right-hand side of (7.92) is

$$-(mh)\int_{h_1}^{h_2} \frac{dh}{h(h-h_a)} = -m_2h_2\int_{h_1}^{h_2} \frac{dh}{h(h-h_a)},$$

which equals

$$-m_2h_2\int_{h_1}^{h_2} \frac{1}{h_a}\left(\frac{1}{(h-h_a)} - \frac{1}{h}\right)dh,$$

where h_a is constant.

Then

$$m_a = -\frac{m_2h_2}{h_a}\ \ell n\left(\frac{h_2-h_a}{h_1-h_a}\ \frac{h_1}{h_2}\right)$$

or

$$-\frac{m_ah_a}{m_2h_2} = \ell n\left(\frac{h_2-h_a}{h_1-h_a}\ \frac{h_1}{h_2}\right). \qquad (7.93)$$

Now the definition for the scavenge ratio λ is

$$\lambda = \frac{m_a}{m'},$$

where

$$m' = \frac{p_sV_T}{RT_s} = \frac{p_aV}{RT_a}$$

or
$$m'RT_a = p_aV.$$

Now
$$m'RT_a = m'\frac{\kappa-1}{\kappa}C_p T_a = m'\frac{\kappa-1}{\kappa}h_a.$$

Hence
$$m'h_a = \frac{\kappa}{\kappa-1}p_aV.$$

Similarly,
$$m_2h_2 = \frac{\kappa}{\kappa-1}p_2V.$$

For a constant pressure process,
$$p_2 = p_a,$$

and we have
$$p_2V = p_aV$$

or
$$m_2h_2 = m'h_a.$$

Hence
$$\lambda = \frac{m_a}{m'} = \frac{m_ah_a}{m'h_a} = \frac{m_ah_a}{m_2h_2}.$$

Equation (7.93) now becomes
$$\lambda = \ell n\left(\frac{h_2-h_a}{h_1-h_a}\frac{h_1}{h_2}\right)$$

or
$$e^{-\lambda} = \left(\frac{h_2-h_a}{h_2}\right)\left(\frac{h_1}{h_1-h_a}\right), \qquad (7.94)$$

$$e^{-\lambda} = \left(1-\frac{h_a}{h_2}\right)\left(\frac{1}{1-\frac{h_a}{h_1}}\right).$$

For ideal gas with constant specific heats,
$$h = C_pT$$

assuming
$$(C_p)_g = (C_p)_a,$$

and we obtain
$$e^{-\lambda} = \left(1-\frac{T_a}{T_2}\right)\left(\frac{1}{1-\frac{T_a}{T_1}}\right)$$

$$\left(1-\frac{T_a}{T_1}\right)e^{-\lambda} = 1-\frac{T_a}{T_2}.$$

Hence
$$T_2 = \frac{T_a}{1-\left(1-\frac{T_a}{T_1}\right)e^{-\lambda}}. \tag{7.95}$$

The temperature T_2 is the temperature at the end of scavenging and T_1 is the temperature at the commencement of scavenging, i.e. at the end of blowdown.

The scavenge process is a mixing process and we can, therefore, use the same methods as outlined earlier for evaluating the scavenge efficiency with variable temperature of the cylinder contents.

Let the percentage by mass of air in the cylinder at any time be Z. Then the mass of air entering the cylinder at any instant of time is dm_a and the mass of air leaving is $Z\, dm_e$, where dm_e is the mass of the mixture of air plus residuals.

The mass of air retained is dm_a' and is given by

$$dm_a' = dm_a - Z\, dm_e,$$

where
$$Z = \frac{m_a'}{m}.$$

Hence
$$m_a' = mZ,$$

$$dm_a' = m\, dZ + Z\, dm.$$

Therefore
$$m\, dZ + Z\, dm = dm_a - Z\, dm_e.$$

Now
$$dm_e = dm_a - dm \quad \text{from the mass balance.}$$

Hence
$$m\, dZ + Z\, dm = dm_a - Z\,(dm_a - dm)$$

or
$$m\, dZ = dm_a(1-Z).$$

Therefore
$$\frac{dm_a}{m} = \frac{dz}{1-Z}.$$

Now from (7.91)
$$\frac{dm_a}{m} = -\frac{dh}{h-h_a}.$$

Hence
$$\int\frac{dZ}{1-Z} = -\frac{dh}{h-h_a}.$$

Once again we use limits (1), (2) corresponding to APO, APC at APO $Z = 0$, at APC $Z = \eta_{sc}$, the scavenge efficiency.

$$\int_0^{\eta_{SC}} \frac{dZ}{1-Z} = -\int_{h_1}^{h_2} \frac{dh}{h-h_a}$$

$$-\ln(1-\eta_{SC}) = -\ln\left(\frac{h_2-h_a}{h_1-h_a}\right)$$

or
$$1-\eta_{SC} = \frac{h_2-h_a}{h_1-h_a}. \tag{7.96}$$

Now from (7.94)
$$e^{-\lambda} = \frac{h_2-h_a}{h_1-h_a}\frac{h_1}{h_2}$$

$$\frac{h_2-h_a}{h_1-h_a} = \frac{h_2}{h_1}e^{-\lambda}.$$

Hence substitute into (7.96)
$$1-\eta_{SC} = \frac{h_2}{h_1}e^{-\lambda} = \frac{T_2}{T_1}e^{-\lambda}$$

or
$$\eta_{SC} = 1-\frac{T_2}{T_1}e^{-\lambda}. \tag{7.97}$$

This is the modified Hopkinson equation for a non-isothermal but adiabatic process ($\dot{Q} = 0$).

The variation of scavenge efficiency η_{SC} with temperature ratio T_2/T_1 and scavenge ratio λ is shown in Fig. 7.18. The temperature ratio T_2/T_1 decreases with increase in engine load. Thus as the engine load increases so the scavenge efficiency increases.

For the special case $T_1 = T_2$ or an "isothermal process" we have $\eta_{SC} = 1-e^{-\lambda}$, but notice although this gives the same answer as the isothermal process in the present case the temperature is constant because there is heat transfer. In the Hopkinson mixing we assumed constant T and made $T_1 = T_2 = T_a$, in this case we should have an adiabatic isothermal process, and we now have an alternative proof to the Hopkinson equation although more laborious.

Now the volumetric efficiency η_V is

$$\eta_V = \left(\frac{p_c}{p_s}\right)\left(\frac{T_s}{T_c}\right).$$

For a constant pressure process we have assumed

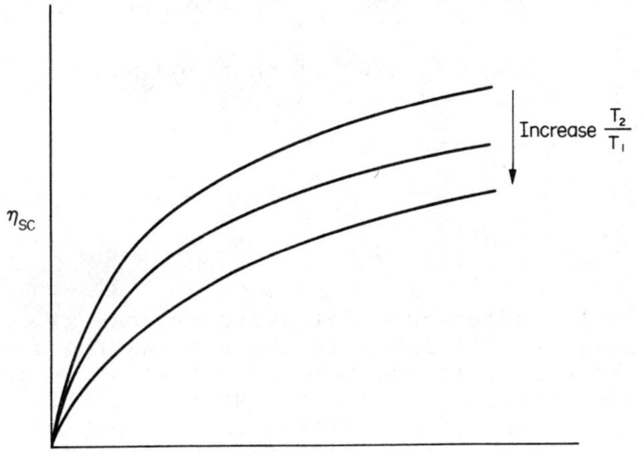

FIG. 7.18. Variation of scavenge efficiency with temperature and scavenge ratio.

$$p_s = p_a = p_c = p.$$

Also $$T_c = T_2, \quad T_s = T_a.$$

Then $$\eta_V = \frac{T_a}{T_2}.$$

Now from (7.95)

$$\frac{T_a}{T_2} = 1-\left(1 - \frac{T_a}{T_1}\right) e^{-\lambda}.$$

Hence $$\eta_V = 1-\left(1 - \frac{T_a}{T_1}\right) e^{-\lambda}$$

$$= 1-e^{-\lambda} + \frac{T_a}{T_1} e^{-\lambda}$$

$$= 1-e^{-\lambda} + \frac{T_a}{T_2}\frac{T_2}{T_1} e^{-\lambda}$$

$$= 1-e^{-\lambda} + \eta_V\left(\frac{T_2}{T_1} e^{-\lambda}\right).$$

Now from (7.97)

$$\frac{T_2}{T_1} e^{-\lambda} = 1 - \eta_{SC}.$$

Hence $$\eta_V = 1-e^{-\lambda} + \eta_V(1-\eta_{SC}),$$

$$\eta_V = 1-e^{-\lambda} + \eta_V - \eta_V\eta_{SC},$$

or $$\eta_V\eta_{SC} = 1-e^{-\lambda}.$$

Hence $$\eta_{CH} = 1-e^{-\lambda}. \qquad (7.98)$$

Thus for a variable temperature adiabatic process at constant pressure the charging efficiency is the same as the charging efficiency for an adiabatic isothermal process. In practice this implies that model tests at constant temperature will give a measure of the corresponding charging efficiency in an engine. This is the theoretical basis for scavenge model testing. A review of a number of methods is given by Benson.[4]

Complex gas dynamic models have been developed for mixing-displacement-short circuiting processes; these have been described elsewhere.[3]

7.5 FLOW PROCESSES IN THE GAS EXCHANGE PERIOD

The flow into and out of the cylinder is controlled by the conditions in the inlet system, the cylinder and the exhaust system. The controlling parameters are the dimensions of the system. In this text we can only examine the simplest flow problems. These will neglect pressure fluctuations in the various parts of the system except in the case of the exhaust blowdown period, when we shall examine the conditions controlling the pressure drop in the cylinder.

We shall first examine the methods for calculating the exhaust valve port areas followed by the corresponding calculations for air ports for two-stroke cycle engines and air valves for four-stroke engines.

7.5.1 Exhaust Valve or Port Area

The major requirement of the exhaust valve (or ports) is to drop the pressure in the cylinder to the air supply pressure as quickly as possible. This corresponds to the blowdown period.

Both exhaust valves and ports will be treated in the same way to evaluate the port or valve areas.

We shall make the following assumptions:

(1) Constant downstream pressure in exhaust pipe for subsonic flow through the ports or valve and critical pressure for choked flow through the ports or valve.

(2) No heat transfer. Thus the expansion is isentropic in the cylinder.

(3) Isentropic expansion through the ports or valves.

We shall first derive a general expression for a variable volume V_c corresponding to a cylinder length L_c and cross-sectional area F_c. We shall solve the resultant differential equation for a constant cylinder volume (Fig. 7.19).

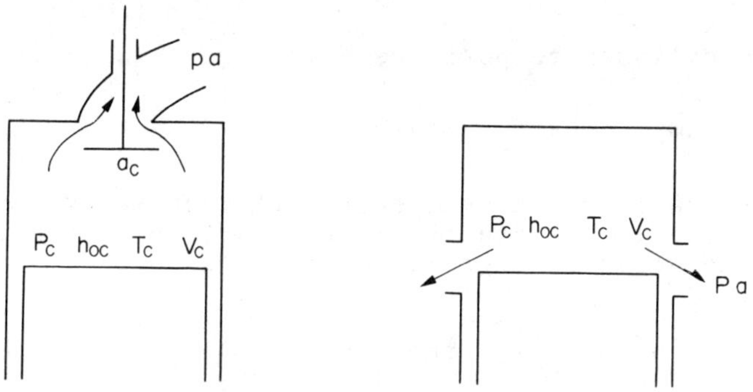

FIG. 7.19. Exhaust valve or port area.

The mass in the cylinder is given by

$$p_c V_c = m_c R T_c$$

or

$$\frac{dm_c}{m_c} = \frac{dp_c}{p_c} + \frac{dV_c}{V_c} - \frac{dT_c}{T_c}. \tag{7.99}$$

Now

$$a_c^2 = \kappa R T_c.$$

Hence

$$m_c = \frac{\kappa p_c V_c}{a_c^2} \tag{7.100}$$

For isentropic expansion in the cylinder and across the ports,

$$\frac{dp_c}{p_c} = \frac{\kappa}{\kappa-1}\frac{dT_c}{T_c} = \frac{2\kappa}{\kappa-1}\frac{da_c}{a_c} \tag{7.101}$$

and

$$\frac{p}{p_c} = \left(\frac{a}{a_c}\right)^{\frac{2\kappa}{\kappa-1}} = x^{\frac{2\kappa}{\kappa-1}} \tag{7.102}$$

Also $$V_c = F_c L_c, \tag{7.103}$$

where F_c is the cylinder cross-sectional area and L_c is the cylinder length.

Substituting (7.100) to (7.103) into (7.99) and rearranging,

$$dm_c = \frac{\kappa p_c}{a_c^2} F_c dL_c + \frac{2\kappa}{\kappa-1} p_c \frac{L_c F_c}{a_c^3} da_c. \tag{7.104}$$

For flow from cylinder to ports we have

$$dm_c = -\phi F_c u \rho \, dt, \tag{7.105}$$

where ϕ is the ratio of the valve or port area to cylinder cross-sectional area, ρ is the density in the port and u the corresponding velocity. The negative sign is for the mass balance in the cylinder.

If we let $b = \frac{2}{\kappa-1}$, then

$$\rho = \rho_c x^b \tag{7.106}$$

from (7.102).

The energy equation (2.83) from the cylinder to the port is

$$h + \frac{u^2}{2} = h_{oc}. \tag{7.107}$$

Now $$h = \frac{a^2}{\kappa-1}, \quad h_{oc} = \frac{a_c^2}{\kappa-1}, \tag{7.108}$$

then (7.107) becomes

$$u = \sqrt{b}\, a_c (1-x^2)^{\frac{1}{2}} \tag{7.109}$$

using (7.103) and (7.108).

It follows from (7.106) and (7.109) that the mass balance in the cylinder (7.105) is

$$dm_c = -\phi F_c a_c \rho_c \sqrt{b}\, (1-x^2)^{\frac{1}{2}} x^b dt. \tag{7.110}$$

Combining (7.104) and (7.110) we have after simplification

$$\frac{dL_c}{\sqrt{b}\, a_c} + \frac{\sqrt{b}.L_c.da_c}{a_c^2} = -\phi(1-x^2)^{\frac{1}{2}}\, x^b dt. \tag{7.111}$$

We can solve this equation analytically for certain set conditions of isentropic index κ. We will first consider a constant volume cylinder ($dL_c = 0$) for subsonic flows. For subsonic flow the pressure in the port p is equal to the constant exhaust pressure p_{exh}. The release pressure at exhaust port opening is p_R.

We define

$$x_R = \frac{a}{a_R} = \left(\frac{p}{p_R}\right)^{\frac{\kappa-1}{2\kappa}} = \left(\frac{p_{exh}}{p_R}\right)^{\frac{\kappa-1}{2\kappa}} = \text{constant.}$$

Now

$$x = \frac{a}{a_c} = \frac{a}{a_R}\frac{a_R}{a_c} = \frac{x_R a_R}{a_c}$$

or

$$a_c x = x_R a_R = \text{constant} \tag{7.112}$$

since

$$a_R = \sqrt{\kappa R T_R}.$$

Hence

$$\frac{da_c}{a_c} + \frac{dx}{x} = 0. \tag{7.113}$$

Substituting (7.112) and (7.113) into (7.111) and setting $dL_c = 0$, we obtain after rearrangement and integrating,

$$\frac{x_R a_R}{L_c}\int \phi\, dt = \sqrt{b}\int_{x_R}^{x} \frac{dx}{x^b(1-x^2)^{\frac{1}{2}}}. \tag{7.114}$$

The analytical solution of the right-hand side of (7.114) is

$$-\sqrt{b}\left[\frac{1}{(b-1)}\frac{(1-x^2)^{\frac{1}{2}}}{x^{b-1}} + \left(\frac{b-2}{b-1}\right) \sec^{b-2}\theta\, d\theta\right], \tag{7.115}$$

where x is equal to $\cos\theta$ and κ is equal to $1 + \frac{2}{b}$.

The solution only applies for fixed values of $\kappa = 1+(2/b)$. For sonic flow through the port the velocity through the ports u equals the local speed of sound a.

The mass balance (7.105) is

$$dm_c = -\phi F_c\, u\, \rho\, dt \tag{7.116}$$

and (7.106) is

$$\rho = \rho_c \left(\frac{2}{\kappa+1}\right)^{\frac{\kappa+1}{2(\kappa-1)}} . \tag{7.117}$$

Let $$e = \frac{\kappa+1}{2(\kappa-1)}.$$

Combining (7.104), (7.116) and (7.117) we obtain the differential equation for sonic flow,

$$\left(\frac{\kappa+1}{2}\right)^e \frac{dL_c}{a_c L_c} + b\left(\frac{\kappa+1}{2}\right)^e \frac{da_c}{a_c^2} = -\frac{\phi_c}{L_c}\,dt. \tag{7.118}$$

For a cylinder of constant length $dL_c = 0$,

$$b\left(\frac{\kappa+1}{2}\right)^e \frac{da_c}{a_c^2} = -\frac{\phi_c}{L_c}\,dt. \tag{7.119}$$

For sonic flow, as above, the port pressure is not constant. If the flow is not choked we let P_{exh} be the constant exhaust pressure and we now define

$$x = \left(\frac{p_{exh}}{p_c}\right)^{\frac{\kappa-1}{2\kappa}} = \left(\frac{p_{exh}}{p_R}\,\frac{p_R}{p_c}\right)^{\frac{\kappa-1}{2\kappa}} = x_R\,\frac{a_R}{a_c}.$$

Hence $$x\,a_c = x_R\,a_R$$

and $$\frac{dx}{x} + \frac{da_c}{a_c} = 0. \tag{7.120}$$

Substituting into (7.119) and integrating,

$$\frac{x_R a_R}{L_c}\int \phi\,dt = b\left(\frac{\kappa+1}{2}\right)^e (x_2 - x_R), \tag{7.121}$$

$$x_2 = \left(\frac{p_{exh}}{p_{c2}}\right)^{\frac{\kappa-1}{2\kappa}} \qquad x_R = \left(\frac{p_{exh}}{p_R}\right)^{\frac{\kappa-1}{2\kappa}}$$

For wholly sonic flow (7.121) applies and for wholly subsonic flow (7.114) applies. For the general case of sonic and subsonic flow both equations must be used. In this case x will correspond to the cylinder pressure at which choked flow ceases (in (7.121)) and is set to x_R in (7.114). The combined equations can then be represented by

$$\frac{x_R a_R}{L_c} \int \phi\, dt = f_1(p_R, p_c). \qquad (7.122)$$

The port area parameter $\int \phi\, dt$ is usually in the form $\int \phi\, d\alpha$ (Fig. 7.20). The two port area parameters are related by the speed N rev/s by

$$\int \phi\, dt = \frac{1}{360N} \int \phi\, d\alpha. \qquad (7.123)$$

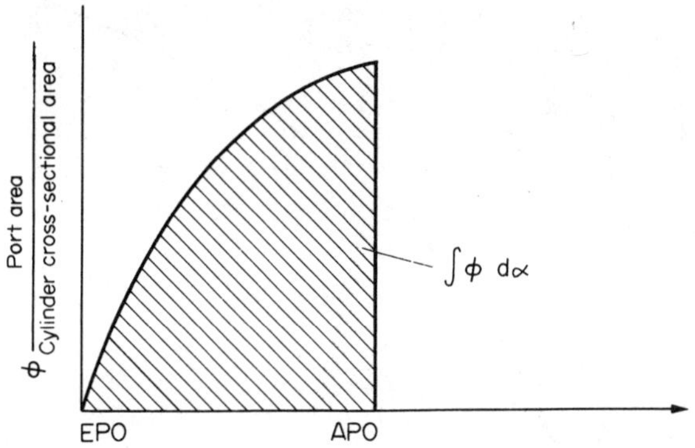

FIG. 7.20. $\int \phi\, d\alpha$ for exhaust port or valve.

We can replace a_R and x_R by T_R, p_R, p_{exh} and (7.122) becomes

$$\frac{T_{cR}}{\left(\dfrac{p_R}{p_{exh}}\right)^{\frac{\kappa-1}{2\kappa}}} \frac{1}{L_c N} \int \phi\, d\alpha = K\, f_2(r_c, r_R), \qquad (7.124)$$

where $$r_c = \frac{p_c}{p_{exh}}, \qquad r_R = \frac{p_R}{p_{exh}}$$

$$K = \text{constant according to units.}$$

Benson[5] has given a number of numerical solutions for $f_2(r_c, r_R)$ for different values of κ. A simple method of applying (7.124) is to replace $f_2(r_c, r_R)$ by $I = f(r_c)$ with $I = 0$ at $r_c = 1.0$. We then obtain one curve for I versus r_c as shown in Fig. 7.21 for $\kappa = 1.4$.

Tabulated values of I are given in Table 7.2 (for $\kappa = 1.4$).

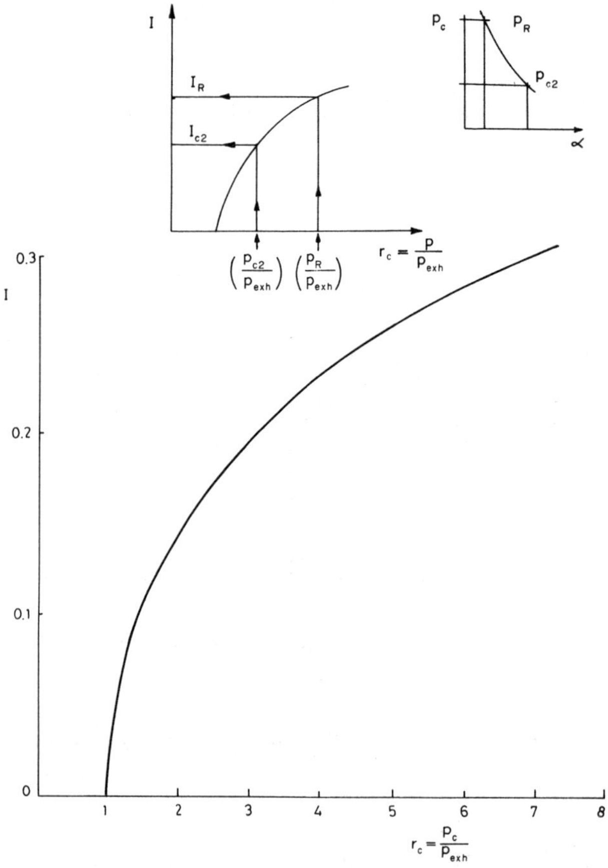

FIG. 7.21. Function I.

TABLE 7.2

r_c	I	r_c	I
1.0	0	2	0.1426
1.073	0.039	3.03	0.198
1.054	0.057	4	0.233
1.242	0.071	5	0.26
1.339	0.082	6.25	0.286
1.446	0.096	7.143	0.302
1.571	0.108		
1.694	0.119		
1.841	0.131		
1.892	0.135		

Equation (7.124) can now be expressed in the form

$$\int \phi \, d\alpha = \frac{L_c N}{\sqrt{T_R}} \left(\frac{p_R}{p_{exh}}\right)^{\frac{\kappa-1}{2\kappa}} K(I_R - I_{c2}), \tag{7.125}$$

where I_R corresponds to $\frac{p_R}{p_{exh}}$ (Fig. 7.21),

I_{c2} corresponds to $\frac{p_{c2}}{p_{exh}}$ (Fig. 7.21).

The constant K is:

$K = 60$, $T_R = {}^oR$, $L_c = ft$, $N = rev/s$.

$K = 146.7$, $T_R = {}^oK$, $L_c = m$, $N = rev/s$.

The application of this curve will be illustrated by a worked example.

Example

Determine $\int \phi \, d\alpha$ for the exhaust ports for the exhaust blowdown of a two-stroke cycle engine. The engine bore is 0.762 m, effective cylinder length during blowdown 2.225 m, release pressure and temperature 4.14 bar and 890°K, exhaust pressure 1.034 bar, speed 2 rev/s.

Solution

$$\frac{p_R}{p_{exh}} = \frac{4.14}{1.034} = 4.0,$$ thus $I_R = 0.233$ from Table 7.2 or Fig. 7.21.

$$\frac{p_c}{p_{exh}} = \frac{1.241}{1.034} = 1.2$$ thus $I_c = 0.065$ from Table 7.2 or Fig. 7.21.

$T_R = 890$, $\sqrt{T_R} = 29.8$, $N = 2$ rev/s.

$\kappa = 1.4$, $\frac{p_R}{p_{exh}} = 1.219$, $L_c = 2.225$ m.

Substituting into (7.125)

$$\int \phi \, d\alpha = \frac{146.7 \times (0.233-0.065) \times 1.219 \times 2.225 \times 2}{29.8}$$

$$\int \phi \, d\alpha = 4.5 \text{ deg.}$$

If the estimate discharge coefficient for the ports = 0.72, then the actual $\int \phi_g d\alpha$ will be

$$\int \phi_g d\alpha = \int \frac{\phi \, d\alpha}{C_d} = \frac{4.5}{0.72} = 6.25.$$

If the blowdown duration is fixed, then by adjusting the port width we can obtain the required value of $\int \phi_g d\alpha$. If the port width is fixed we can obtain the required $\int \phi_g d\alpha$ by adjusting the air port opening (APO). If the APO is fixed, the port width is fixed, then we adjust the EPO until the required $\int \phi_g d\alpha$ is obtained, checking that the effective cylinder length is the same as that used in the calculation (Fig. 7.22).

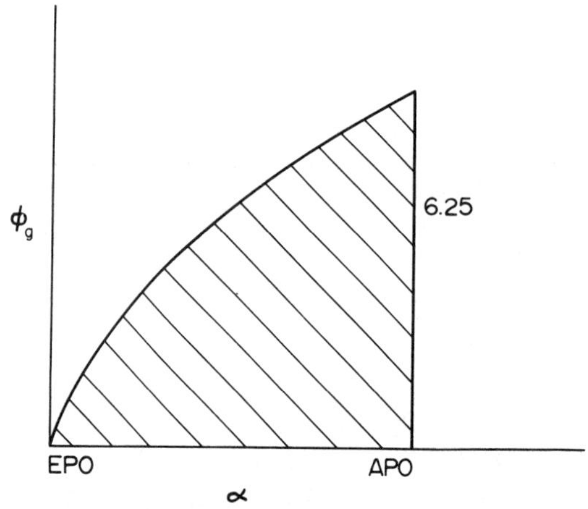

FIG. 7.22. Geometric port area for exhaust blowdown.

In practice the variable exhaust pressure during subsonic flow in the ports influences the blowdown period. It is then necessary to increase $\int \phi_g d\alpha$. This can be allowed for by suitable selection of a discharge coefficient C_d.[6].

7.5.2 Air Port Area in Two-stroke Cycle Engine

If the instantaneous port area is A, the air supply pressure p_a, temperature T_a, then for isentropic flow through the port to cylinder pressure p_c the instantaneous mass flow rate $\dot{m}$ is, from Keenan and Kaye,[7]

$$\dot{m} = \frac{A\,p_a}{\sqrt{T_a}}\,f(r), \tag{7.126}$$

where

$$f(r) = \left[\frac{2w_m}{R_{mol}}\left(\frac{\kappa}{\kappa-1}\right)\right]^{\frac{1}{2}} (r)^{\frac{1}{\kappa}} \left[1-(r)^{\frac{\kappa-1}{\kappa}}\right]^{\frac{1}{2}}, \tag{7.127}$$

$$r = \frac{p_c}{p_a} = \text{pressure ratio across port,}$$

$$w_m = \text{molecular weight of air.}$$

The total mass flow of air supplied per cycle m_s is

$$m_s = \int_{APO}^{APC} \frac{p_a}{\sqrt{T_a}}\,f(r)\,A\,dt \tag{7.128}$$

For constant upstream pressure and temperature and constant cylinder pressure the air supplied per cycle m_s is then

$$m_s = \frac{p_a}{\sqrt{T_a}}\,f(r) \int_{APO}^{APC} A\,dt \tag{7.129}$$

or in terms of the port area-crankangle diagram (Fig. 7.23) $\int A\,d\alpha$ and engine speed N rev/s.

$$m_s = \frac{p_a}{\sqrt{T_a}}\,\frac{f(r)}{360N} \int_{APO}^{APC} A\,d\alpha \tag{7.130}$$

The effective area A is related to the actual geometric port area A_g by the coefficient of discharge C_d,

$$A_g = \frac{A}{C_d}.$$

If an average discharge coefficient is used, then

$$m_s = \frac{p_a}{\sqrt{T_a}}\,\frac{f(r)C_d}{360N} \int_{APO}^{APC} A_g\,d\alpha. \tag{7.131}$$

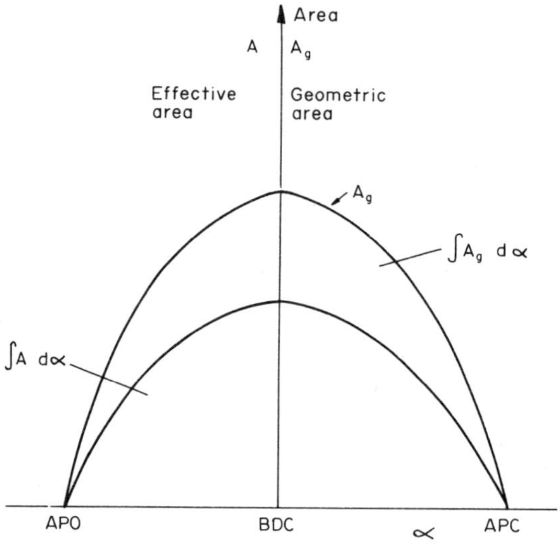

FIG. 7.23. Air port area crankangle diagram.

The mass flow rate supplied to the engine per second $\dot{m}_s$ is given by

$$\dot{m}_s = m_s N. \tag{7.132}$$

It follows, therefore, that the port area times crankangle parameter is

$$\int A_g d\alpha = \dot{m}_s \left(\frac{\sqrt{T_a}}{p_a f(r)} \right) \frac{360}{C_d} \tag{7.133}$$

For a given boost pressure p_a, temperature T_a, pressure ratio across the ports r and mass flow rate $\dot{m}_s$, the port area times angle parameter can be determined for the port.

Normal practice is to compute the port area on the basis of the scavenge ratio λ. This is defined as

$$= \frac{\dot{m}_s}{m'},$$

where $$m' = \frac{p_a V}{R T_a} N$$

and V is the trapped volume at air ports close.

Then $$\int A_g d\alpha = 360 \lambda m' \left(\frac{\sqrt{T_a}}{p_a f(r)} \right) \frac{N}{C_d}$$

$$\int A_g d\alpha = \left(\frac{360\lambda V}{\sqrt{T_a}}\right)\left(\frac{1}{R_f(r)}\right)\frac{N}{C_d}. \tag{7.134}$$

The expression gives the required port area times crankangle $\int A_g d\alpha$ for a given scavenge ratio λ, boost pressure to cylinder pressure ratio r, cylinder volume V, speed N and inlet temperature T_a.

Once a design has been fixed the parameter $\int A_g d\alpha$ is fixed, then for constant scavenge ratio λ, inlet temperature T_a and discharge coefficient C_d, the parameter

$$\left(\frac{\int A_g d\alpha}{360\lambda VC_d}\right)\left(\sqrt{T_a}\,R\right) \text{ is a constant, say, C,}$$

then

$$f(r) = CN$$

or the pressure ratio is a function of the engine speed as shown in Fig. 7.24.

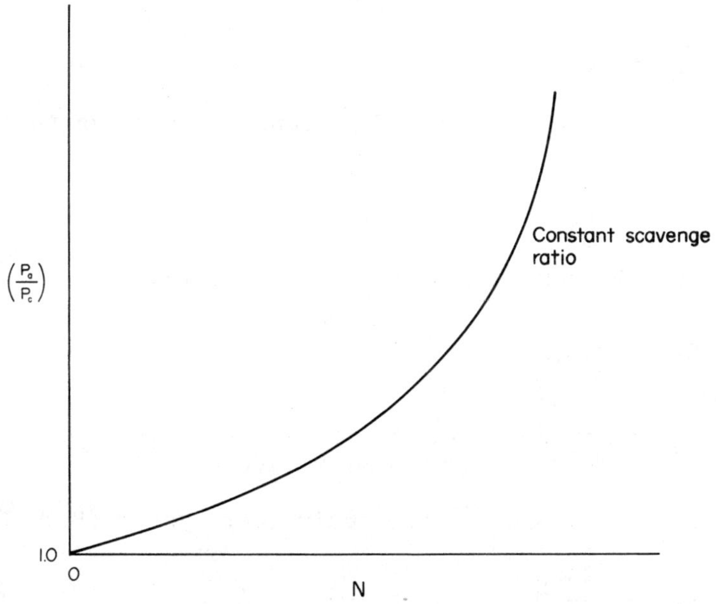

FIG. 7.24. Variation of pressure ratio across air port with speed for constant scavenge ratio.

Such a characteristic is called the engine flow characteristic and the compressor must supply air to follow this line if the scavenge ratio—and hence the scavenge efficiency—is to remain constant over the speed range. A discussion on this characteristic is given by Benson and Horlock[8] and we shall refer to this latter in Chapter 10.

7.5.3 Reduced Port Area

We have up to now considered constant cylinder pressure and air pressure over the scavenge period in two-stroke engines. We have also considered that the cylinder pressure is not influenced by the through flow. In practice both the air and exhaust ports are open at the same time. To allow for the restrictive effect of the exhaust ports or valves we use the concept of the reduced port area.

For this purpose we assume that the flow is incompressible across both sets of ports (or valve).

Let A_s be the instantaneous air port area and A_e be the instantaneous exhaust port or valve area.

Then we assume a model corresponding to two orifices in series (Fig. 7.25b).

We shall make the following assumptions:

(1) $\rho_a = \rho_c = \rho_e$;

(2) the volume flow rate $\dot{V}$ is constant through the system.

Now
$$\dot{V} = \frac{\dot{m}}{\rho_a} = A_s u_s = A_e u_e$$

and from Bernoulli for incompressible flow (2.81)

$$\frac{\Delta p}{\rho} = \frac{u^2}{2}.$$

Hence,
$$u_s^2 = \frac{2\Delta p_s}{\rho_s}, \qquad u_e^2 = \frac{2\Delta p_e}{\rho_e}.$$

Now let subscript E refer to the equivalent system (Fig. 7.25c).

$$u_E = \frac{2\Delta p_E}{\rho_E}, \qquad \dot{V} = \frac{\dot{m}}{\rho_a} = A_E u_E$$

and
$$\Delta p_E = \Delta p_s + \Delta p_e.$$

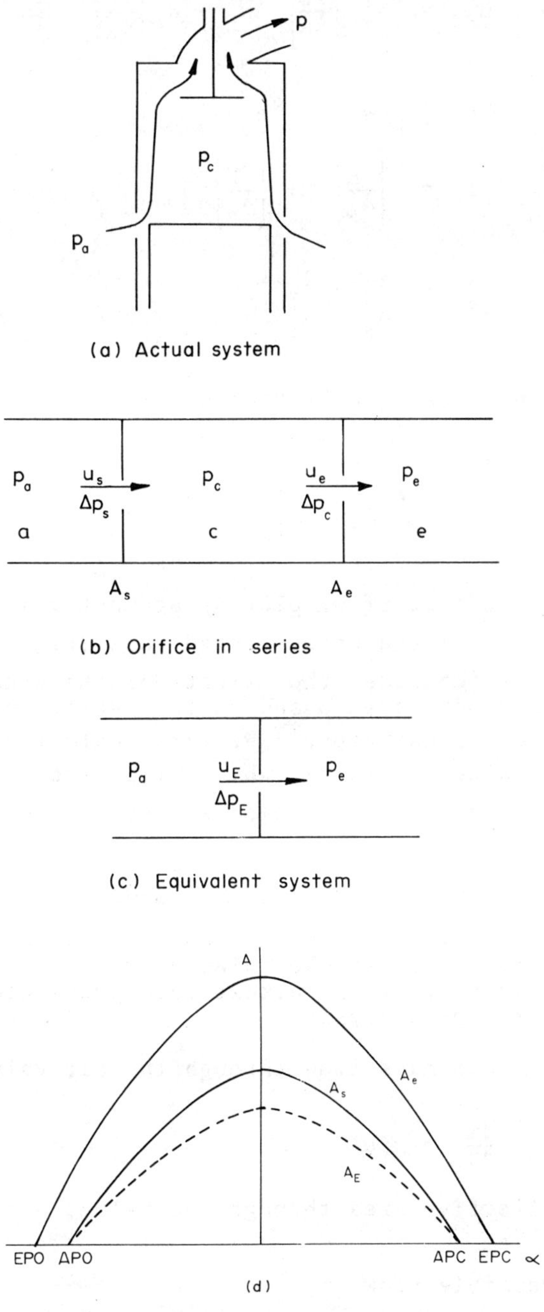

FIG. 7.25. Reduced port area.

Therefore $\rho_E u_E^2 = \rho_s u_s^2 + \rho_e u_e^2.$

Now $$\rho_E u_E^2 = \rho_s \left(\frac{A_E}{A_S}\right)^2 u_E^2 + \rho_e \left(\frac{A_E}{A_e}\right)^2 u_E^2.$$

Since $$\rho_E = \rho_s = \rho_e$$

then $$1 = \left(\frac{A_E}{A_s}\right)^2 + \left(\frac{A_E}{A_e}\right)^2$$

or $$\frac{1}{A_E^2} = \frac{1}{A_s^2} + \frac{1}{A_e^2}$$

The equivalent area A_E will then be

$$A_E = \sqrt{\frac{1}{\frac{1}{A_s^2} + \frac{1}{A_e^2}}}\,.$$

It will be seen that if we plot A_E against α it will give a smaller $\int A\,d\alpha$ than for the unrestricted case (Fig. 7.25d). It is usual, therefore, to consider the effects of the exhaust ports when calculating the air port area based on the overall pressure drop from air supply to exhaust pipe. We first calculate the $\int A_E\,d\alpha$, then we select a value of $\int A\,d\alpha$, which will give A_s-α plot, then check whether together with A_e-α plot we obtain the required $\int A_E\,d\alpha$.

7.5.4 Air Valve Area for Four-stroke Engine

The previous analysis of the air ports of a two-stroke engine could be applied to a four-stroke inlet valve. In practice, the flow through the inlet valve is almost incompressible and we may, therefore, simplify the analysis.

The instantaneous mass flow through the air valve is

$$\frac{dm}{dt} = \rho u A, \tag{7.136}$$

where A is the effective area through the valve, ρ is the density and u the velocity.

For incompressible flow,

$$\frac{dp}{\rho} = \frac{u^2}{2}$$ from Bernoullis' equation (2.80),

where dp is the pressure drop across the valve.

Hence $$u = \sqrt{\frac{2dp}{\rho}},$$

$$\frac{dm}{dt} = \rho\sqrt{\frac{2dp}{\rho}}\,A = A\sqrt{2\rho\,dp}.$$

Now $$\rho \doteq \frac{2p_a}{RT_a} = \text{constant},$$

where suffix a refers to inlet conditions.

Hence $$\frac{dm}{dt} = CA\sqrt{dp}, \qquad (7.137)$$

$$C = \sqrt{\frac{2p_a}{RT_a}}$$

and $$m = C\sqrt{dp}\int_{AVO}^{AVC} A\,dt, \qquad (7.138)$$

where dp is assumed constant.

As before, $$dt = \frac{d\alpha}{360N},$$

where N is rev/s.

In place of $\int A\,d\alpha$ we can write $\bar{A}\alpha_o$, where $\bar{A}$ is the mean valve area, α_o is the duration of opening and

$$\bar{A} = \frac{1}{\alpha_o}\int A\,d\alpha. \qquad (7.139)$$

The mean valve area $\bar{A}$ is shown in Fig. 7.26.

The mass flow per cycle m is then

$$m = \frac{C}{360N}\sqrt{dp}\;\bar{A}\,\alpha_o. \qquad (7.140)$$

Now the mass flow per cycle is approximately equal to the density ρ_a times the swept volume V_S.

The swept volume V_S = stroke x cross-sectional area of cylinder = SF_c

or $$m = \rho_a SF_c.$$

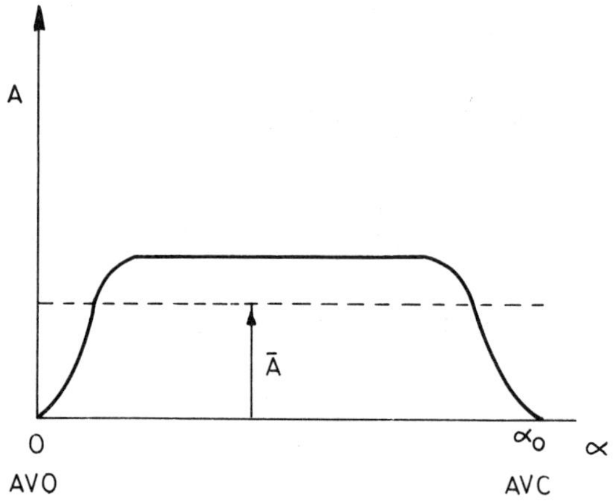

FIG. 7.26. Air valve area/time diagram.

Substitute for m in (7.140) and rearrange, we then obtain the pressure drop across the valve as

$$dp = \left(\frac{SN}{(\bar{A}/F_c)\alpha_o}\right)^2 \left(\frac{360}{C}\rho_a\right) \qquad (7.141)$$

The second term in (7.141) is a constant.

Hence we see that

$$\left(\begin{array}{l}\text{The pressure drop}\\ \text{across the valve}\end{array}\right) \propto \left(\frac{(\text{stroke})^2(\text{speed})^2}{(\text{angle of opening})^2(\text{mean valve area/cylinder cross-sectional area})^2}\right)$$

$$\propto \left(\frac{(\text{piston speed})^2}{(\text{angle of opening})^2(\text{mean valve area/cylinder cross-sectional area})^2}\right)$$

Therefore the pressure drop across the valve:

(i) increases with increase in piston speed squared;

(ii) decreases with increase in valve opening angle squared and increase in valve area squared.

The volumetric efficiency is given by

$$\left(\frac{p_c}{p_a}\right)\left(\frac{T_a}{T_c}\right),$$

where p_c is the cylinder pressure and p_a is the air inlet pressure.

Now $$p_c = p_a - d_p.$$

Hence, neglecting temperature effects,

$$\eta_V \propto \left(\frac{p_a - dp}{p_a}\right) \propto \left(1 - \frac{dp}{p_a}\right).$$

Thus the volumetric efficiency decreases with increase in piston speed and increases with increase in valve opening area and timing.

Finally, it will be observed that the ratio

$$\frac{\text{mean valve area}}{\text{cylinder cross-section area}}, \text{ often called the } \frac{\text{valve area}}{\text{piston area}} \text{ ratio,}$$

is an important parameter. The greater this value the higher the volumetric efficiency.

7.6 SPARK IGNITION GASOLENE ENGINE INTAKE SYSTEM—CARBURETTOR

The fuel/air mixture is prepared in the intake system of a spark ignition engine in the carburettor. Modern carburettors are extremely complex, but we can describe their general functions in terms of simple carburettors. In Figs. 7.27 and 7.28 an elementary updraught carburettor and a downdraught carburettor are shown. The intake pipe consists essentially of a venturi tube called a choke tube at the minimum area of which is a jet pipe connected to the fuel supply. The level of the fuel supply relative to the tip of the jet pipe is kept constant by a float. The air flow rate to the engine is controlled by a throttle downstream of the choke tube. The pressure difference between the fuel float chamber, normally at the outside pressure, and the choke tube causes the fuel to flow along the jet tube and to be sprayed into the air stream as a series of droplets. The size of the droplets will depend on the spray dimensions, the relative velocity of the air, the surface tension, viscosity and volatility of the fuel. Excepting under part load conditions, the fuel is not completely vaporized in the intake manifold and vaporization is completed in the engine cylinder. The smaller the drop size, the greater the tendency to vaporization. At low air flow ratios, corresponding to engine idling conditions, the pressure difference between the float chamber and the choke tube is inadequate to cause the fuel to flow due to fuel viscosity and surface tension. A second jet is therefore required. This jet is located at the tip of the throttle valve so that the depression at this point is adequate to cause the fuel to flow. Under cold conditions, when starting, the cylinder wall temperature is too low to assist the final stages of vaporization of the fuel in the cylinder, so that the fuel vapour/air ratio is not adequate for ignition. For these conditions a second butterfly valve called a choke is fitted upstream of the jet. This valve is almost closed during starting, further restricting the air flow, thus effectively increasing the fuel vapour/air ratio in the cylinder to aid ignition.

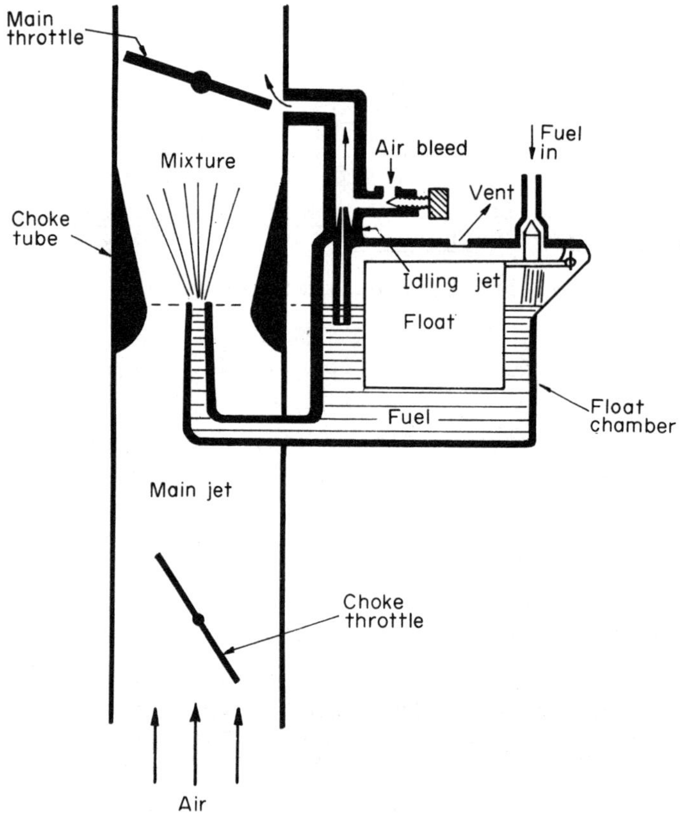

FIG. 7.27. Updraught carburettor.

We can model the carburettor by a simple analysis. Let A_c be the cross-sectional area of the choke tube and A_j the cross-sectional area of the fuel jet. The pressure difference in the fuel system can be divided into two parts; one part (dp_1) corresponding to the friction losses in the fuel passage plus the surface tension force at the jet tip, and the second part (dp_2) the pressure difference between the jet tip and the choke tube. The total pressure difference between the float chamber and the choke throat is then

$$dp = dp_1 + dp_2. \tag{7.142}$$

The fuel jet velocity u_f, from Bernoullis equation (2.81) for incompressible fluid, is

$$u_f = \sqrt{\frac{2dp_2}{\rho_f}}, \tag{7.143}$$

where ρ_f is the density of the fuel. The fuel flow rate $\dot{m}_f$ is then

$$\dot{m}_f = \rho_f u_f A_j = A_j \sqrt{2dp_2\rho_f},$$

$$\dot{m}_f = A_j \sqrt{2\rho_f(dp-dp_1)}. \tag{7.144}$$

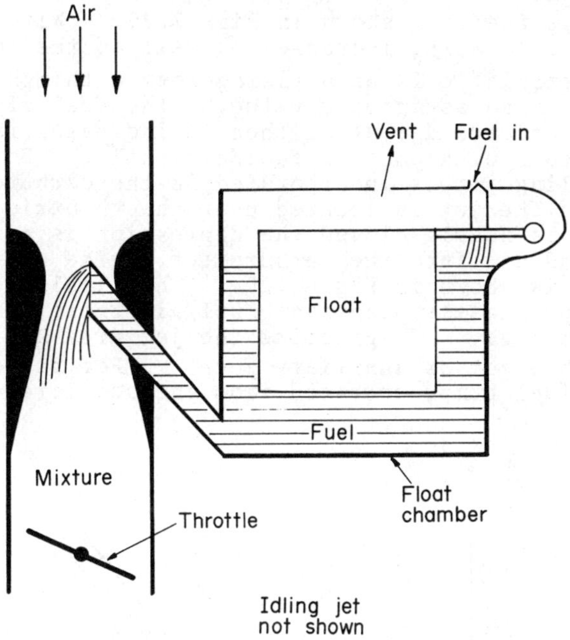

FIG. 7.28. Downdraught carburettor.

For incompressible flow the air velocity through the choke tube u_a, due to a pressure difference dp_a, is, from (2.81),

$$u_a = \sqrt{\frac{2dp_a}{\rho_a}},$$

and the air flow rate $\dot{m}_a$ is

$$\dot{m}_a = \rho_a u_a A_c = A_c \sqrt{2dp_a \rho_a} \ . \quad (7.145)$$

The float chamber is normally vented to atmosphere and $dp = dp_a$, then the fuel/air ratio in the carburettor is

$$\frac{\dot{m}_f}{\dot{m}_a} = \left(\frac{A_j}{A_c}\right) \left(\frac{\rho_f}{\rho_a}\right)^{\frac{1}{2}} \left(\frac{dp-dp_1}{dp}\right)^{\frac{1}{2}} \quad (7.146)$$

$$\frac{\dot{m}_f}{\dot{m}_a} = \left(\frac{A_j}{A_c}\right) \left(\frac{\rho_f}{\rho_a}\right)^{\frac{1}{2}} \left(1- \frac{dp_1}{dp}\right)^{\frac{1}{2}} ; \quad (7.147)$$

for $dp_1/dp = 1.0$ there is no flow.

For constant area ratio A_j/A_c the characteristics of a simple carburettor have the form shown in Fig. 7.29. With increasing mass flow the ratio ρ_f/ρ_a increases as well as the term $(1-(dp_1/dp))$. Thus the characteristic is of a rising form although, as will be seen, it tends to an asymptotic value. The fuel flow does not start until dp exceeds dp_1 at A, then it increases rapidly to B corresponding to stoichiometric fuel/air ratio. For starting purposes an idling jet is incorporated in the carburettor as shown in Fig. 7.27. The jet is located near the throttle valve so that when the valve is nearly closed the depression is exceedingly large, forcing fuel and air into the carburettor. The characteristic of the idling jet is shown in Fig. 7.29. The simple carburettor operates at approximately constant fuel/air ratio over a wide range of throttle settings. In practice the jet area A_j is augmented by either a variable jet or auxiliary jets. For rapid load application a fuel pump, operated from the accelerator link gear, is normally used.

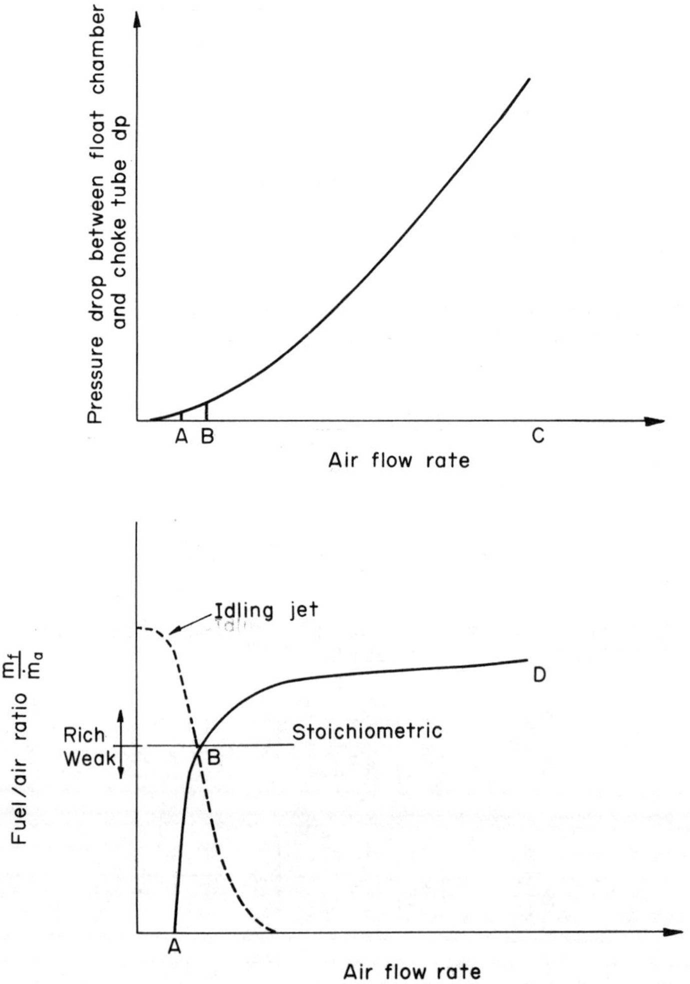

FIG. 7.29. Simple carburettor flow characteristics.

After passing through the choke tube the pressure drop across the choke tube is nearly completely recovered; however, the pressure drop across the throttle valve is not. The pressure loss Δp across a carburettor can be expressed in the form (Benson and Sierens)[9]

$$\Delta p = f \frac{\rho u^2}{2}$$

where f is a constant depending on throttle setting and mach number.

7.7 NON-STEADY FLOW WAVE ACTION

In our discussion of the gas exchange process we have assumed that the pressure in the inlet and exhaust systems was either constant or independent of the cylinder pressure. This is not the case in practice. The release of high pressure gas from the cylinder through the exhaust valves or ports sets in motion pressure waves in the exhaust pipe which are propagated at the local speed of sound relative to the gas velocities. These pressure waves interact with branches in the pipe and at the pipe end. The interactions cause pressure waves to be reflected back to the exhausting cylinder. In multi-cylinder engines, if there are several cylinders exhausting at the same time, there may be interference with the gas exchange process in these cylinders due to the transmitted and reflected waves. The pressure waves may aid or retard the gas exchange process. When they aid the process the exhaust system is said to be tuned. Methods to predict the flows in the engine due to the non-steady flow are now well established. They are, however, fairly complicated. A simplified outline will be given in Chapter 10 (Section 10.8.2).

The variable inlet flow rate to the cylinder in the inlet valve or ports will cause expansion waves to be propagated into the inlet manifold. These expansion waves will be reflected at the open end causing positive pressure waves to be propagated towards the cylinder. If the timing of these waves is well arranged, then the positive pressure wave will cause the trapped pressure to be raised above the nominal inlet pressure. When a design achieves this objective the process is called induction ramming. The same methods may be applied for studying inlet pressure waves as exhaust pressure waves.

In some engines the inlet air supply is through a reciprocating pump or the crankcase is used as a pump. In these cases the inlet manifold pressure will not be constant. In Fig. 7.30 some results are shown for a small crankcase scavenged two-stroke engine. The air/fuel mixture is drawn into the crankcase and compressed as the piston moves downwards. At the appropriate time a transfer port is opened by the piston and the charge passes into the cylinder. The fluctuating pressures are clearly seen.

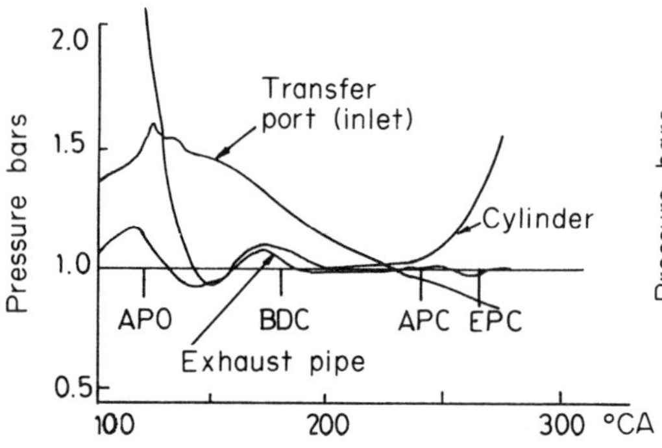

(a) Pressure diagrams at 3000 rev/min

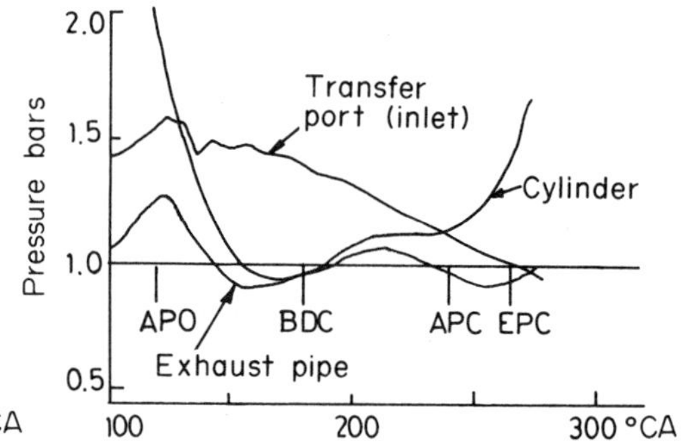

(b) Pressure diagrams at 5850 rev/min

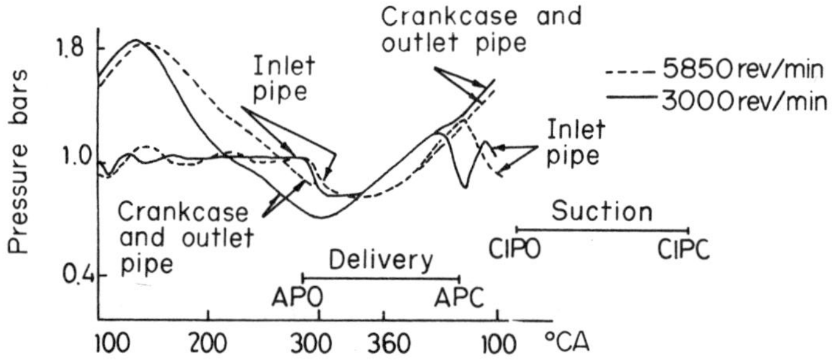

(c) Crankcase pressure-time diagram

FIG. 7.30. Pressure diagrams in crankcase compression engine.

REFERENCES

1. Ricardo, H.R. and Hempson, J.G.G., The High-speed Internal Combustion Engine, 5th edn., Blackie, 1968.

2. Hopkinson, B., The charging of two cycle internal combustion engines, Trans. NE Coast Instn. Engrs. Shipbuilders 30, 333 (1913-14).

3. Benson, R.S. and Brandham, P.J., A method for obtaining a quantitative assessment of the influence of charge efficiency on two-stroke engine performance, Int. J. Mech. Sci. 11, 303 (1969).

4. Benson, R.S., A survey of research on the scavenging of two-stroke cycle engines, BSRA Report No. 90 (1952).

5. Benson, R.S., A method for calculating the exhaust-port area for two-stroke cycle engines, Jl. R. Aeronaut. Soc. 61, 127 (1957).

6. Benson, R.S., The influence of exhaust belt design on the discharge process of a two-stroke cycle engine, Proc. Instn. Mech. Engrs. 174, 713 (1960).

7. Keenan, J.H. and Kaye, J., Gas Tables, John-Wiley, New York, 1948.

8. Benson, R.S. and Horlock, J.H., The matching of two-stroke engines and turbochargers, Proc. 6th Int. Cong. CIMAC, p. 464 (1962).

9. Benson, R.S., Baruah, P.C. and Sierens, R., Steady and non-steady flow in a simple carburettor, Proc. Instn. Mech. Engrs. 188, 53/74 (1974).

Chapter 8

Compression Ignition Engine Cycle Calculations

Notation

a	Annand equation coefficient
act	constant (activation energy) in reaction rate equation
a_i	number of mols of species i at end of step
A_s	exposed surface area of combustion chamber
A/F	air/fuel ratio
$(A/F)_{st}$	stoichiometric air/fuel ratio
b	Annand equation coefficient
b_i	number of mols of species i at beginning of step
C	Annand equation coefficient
CA	percentage of weight of carbon in fuel
CR	compression ratio
C_v	specific heat at constant volume
D	cylinder bore
dM_f	number of mols of fuel burnt in step
e	specific internal energy
E	absolute internal energy of cylinder contents
ΔE_o	heat of reaction of unit molal quantity of fuel at absolute zero
HA	percentage by weight of hydrogen in fuel
k	thermal conductivity
K	preparation rate constant
K'	reaction rate constant
ℓ	connecting rod length
m	number of hydrogen atoms in equivalent hydrocarbon fuel
M	number of mols
M_i	mass of fuel injected
M_u	mass of fuel in cylinder and unprepared
n	number of carbon atoms in equivalent hydrocarbon fuel
n	ratio connecting rod length to half stroke
N	engine speed (revolutions per second)
p	pressure
p_m	indicated mean effective pressure
p_{O_2}	partial pressure of oxygen
p	preparation rate
Q	heat transfer
q_{VS}	lower molal calorific value
Q_{VS}	lower heat of reaction at constant volume
R	reaction rate
R_{mol}	universal gas constant
Re	Reynolds number
S	stroke
t	time
T	temperature
T_c	mean temperature of cylinder contents
T_w	cylinder wall temperature
V	volume
V_c	clearance volume
V_s	swept volume
V_p	mean piston speed
w	total number of mols of fuel
W	work
W_m	molecular weight of fuel
x	index constant in preparation equation
x	proportion of total fuel injected
x_f	proportion of total fuel injected
α	crankangle
μ	viscosity
η_{TH}	indicated thermal efficiency
ρ	density

8.1 INTRODUCTION

When assessing the performance of any engine, whether naturally aspirated or pressure charged, the most important processes involved are those taking place in the engine cylinder, that is the compression, combustion and expansion processes, with the proviso that these can only take place correctly if there is sufficient trapped air for the combustion of the fuel.

The assumption that this statement is valid may often be used to obtain major trends due to many significant variables such as maximum cylinder pressure, compression ratio, trapped pressure and temperature or pattern of combustion, without consideration of the manifold conditions necessarily associated with them, but always remembering that there are such manifold conditions to allow for.

The basic thermodynamics used in these calculations are as already described in Chapter 2. It may also be noted that although spark ignition engine combustion cycles are dealt with separately (in Chapter 9), there is no fundamental thermodynamic difference between spark ignition and compression ignition cycle calculations. The differences are predominantly due to fluid dynamics and mixing characteristics and flame travel effects, which result in different calculation models. The development of stratified charge combustion systems for spark ignition engines and of multi-zone models for diesel combustion has reduced the once wide disparity between the two systems but not sufficiently to overthrow the traditional separate approach.

The straightforward spark ignition combustion system is based upon a homogeneous gas/air combustible mixture through which a flame propagates from the spark. This automatically introduces the concept of two very different zones—of unburnt reactants and of products of combustion—separated by the flame front. The diesel engine combustion system is very much more complicated than this, but its very complication led to the use of a simple model for calculation purposes, based primarily upon insufficient knowledge to justify anything better. This model assumes that conditions—pressure, temperature, composition—are uniform throughout the cylinder contents at any instant of time. Probably most diesel engine cycle calculations used commercially have been, and still are, based upon such a single-zone model. Multi-zone models are developments from it. So it is appropriate to deal with it at some length.

We shall first discuss the ideal combustion cycle and follow with the more realistic cycles. Since compression ignition engines operate with a very much weaker mixture than spark ignition engines, we usually neglect dissociation. This makes the thermodynamic analysis much simpler than for spark ignition engine cycle calculations. To reduce duplication the combustion reactions are developed in Chapter 9 and only stated in this chapter. Whilst it would be possible to present the fuel for compression ignition engines in terms of the constituent hydrocarbons, the fuel composition is usually more complex than for a spark ignition engine fuel. It is usual therefore to specify the composition of the fuel in percentage by weight. In this chapter we shall consider the fuel to be formed of carbon and hydrogen only. To facilitate calculations we set up an equivalent hydrocarbon C_nH_m to represent the fuel. If CA is the percentage by weight of carbon in the fuel then the number of carbon atoms n is equal to CA/12, and if HA is the percentage by weight of hydrogen in the fuel then the number of hydrogen atoms m is equal to HA.

In a compression ignition engine, fuel is continuously injected into the cylinder and the thermodynamics of the combustion process are slightly different from the spark ignition engine process. We shall start our development of the thermodynamics of compression ignition engine cycles by discussing the combustion process before commencing on the cycle details.

8.2 THERMODYNAMICS OF COMBUSTION PROCESS

In Chapter 3 the combustion processes in the air standard cycles were represented by heat transfer processes. In the somewhat more realistic compression ignition engine cycle calculations being considered, this notion is still retained and the combustion model is called a heat release model. Thermodynamically, heat is not released during combustion; the temperature rise is due to the change in composition of the initial reactants—fuel and air—to the products of combustion in an exothermic reaction. The thermodynamics of these processes have been outlined in Chapter 2. In the case of the compression ignition engine, fuel is injected into the cylinder continuously and, as explained in Chapter 4, there is a delay period before the fuel is burnt. In an ideal cycle we assume that the combustion is instantaneous. However, in the more realistic cycle the combustion of fuel in the cylinder takes place in the presence of the products of earlier combustion so that the simple thermodynamics of the combustion models described in Chapter 2 have to be modified to allow for the different nature of the combustion process from the conventional thermodynamic model.

The active constituents participating in the combustion process in the cylinder are the fuel C_nH_m and oxygen according to the reaction (without dissociation).

$$C_n H_m + \left(n + \frac{m}{4}\right) O_2 = nCO_2 + \frac{m}{2} H_2O. \quad (8.1)$$

If dM_f mols of fuel are burnt in a small time step dt, then the mols of fuel in the cylinder will decrease by dM_f, the mols of oxygen decrease by $(n + (m/4)).dM_f$ and the mols of carbon dioxide and water vapour will increase by $n\ dM_f$ and $(m/2)dM_f$ respectively. At any time during the combustion process in the cylinder there will be fuel, oxygen, nitrogen, carbon dioxide and water vapour present. If we use the symbol b to represent the mols at the beginning of a time step dt and a the mols at the end of a time step (this notation being consistent with that used in Chapter 9), then at time t_1 we shall have in the cylinder the species

$$b_1CO + b_2H\,O + b_3O + b_4N + b_5C_nH_m, \tag{8.2}$$

and at time $t_2 = t_1+dt$,

$$a_1CO_2 + a_2H_2O + a_3O_2 + a_4N_2 + a_5C_nH_m. \tag{8.3}$$

If dM_f mols of fuel C_nH_m are burnt in the time step dt, the a and b mols will be related by

$$\left.\begin{aligned} a_1 &= b_1 + n\,dM_f, & a_2 &= b_2 + \frac{m}{2}\,dM_f, \\ a_3 &= b_3 - \left(n + \frac{m}{4}\right)dM_f, & a_4 &= b_4,\ a_5 = b_5 - dM_f. \end{aligned}\right\} \tag{8.4}$$

The <u>absolute</u> energy E_1 of the cylinder contents at time t_1 is from equation (2.98)

$$E_1 = b_1e_{CO_2} + b_2e_{H_2O} + b_3e_{O_2} + b_4e_{N_2} + b_5e_{C_nH_m}$$

or

$$\begin{aligned} E_1 = {} & b_1(e_o)_{CO_2} + b_2(e_o)_{H_2O} + b_3(e_o)_{O_2} + b_4(e_o)_{N_2} \\ & + b_5(e_o)_{C_nH_m} + b_1e(T_1)_{CO_2} + b_2e(T_1)_{H_2O} + b_3e(T_1)_{O_2} \\ & + b_4e(T_1)_{N_2} + b_5e(T_1)_{C_nH_m} \end{aligned} \tag{8.5}$$

and the corresponding <u>absolute</u> internal energy at time t_2 is

$$\begin{aligned} E_2 = {} & a_1(e_o)_{CO_2} + a_2(e_o)_{H_2O} + a_3(e_o)_{O_2} + a_4(e_o)_{N_2} \\ & + a_5(e_o)_{C_nH_m} + a_1e(T_2)_{CO_2} + a_2e(T_2)_{H_2O} + a_3e(T_2)_{O_2} \\ & + a_4e(T_2)_N + a_5e(T_2)_{C_nH_m}. \end{aligned} \tag{8.6}$$

The first law of thermodynamics for the combustion process is in symbolic form

$$dQ - dW = E_2 - E_1.$$

If we let the various internal energy terms be represented by simple functions, namely

$$E(T_2) = a_1e(T_2)_{CO_2} + a_2e(T_2)_{H_2O} + a_3e(T_2)_{O_2} + a_4e(T_2)_{N_2} + a_5e(T_2)_{C_nH_m}, \quad (8.7)$$

$$E(T_1) = b_1e(T_1)_{CO_2} + b_2e(T_1)_{H_2O} + b_3e(T_1)_{O_2} + b_4e(T_1)_{N_2} + b_5e(T_1)_{C_nH_m}, \quad (8.8)$$

$$E_2(T_s) = a_1e(T_s)_{CO_2} + a_2e(T_s)_{H_2O} + a_3e(T_s)_{O_2} + a_4e(T_s)_{N_2} + a_5e(T_s)_{C_nH_m}, \quad (8.9)$$

$$E_1(T_s) = b_1e(T_s)_{CO_2} + b_2e(T_s)_{H_2O} + b_3e(T_s)_{O_2} + b_4e(T_s)_{N_2} + b_5e(T_s)_{C_nH_m}, \quad (8.10)$$

and note from (2.121) that for (8.1) the heat of reaction at absolute zero ΔE_o is given by

$$\Delta E_o = n\left(e_o\right)_{CO_2} + \frac{m}{2}\left(e_o\right)_{H_2O} - \left(e_o\right)_{C_nH_m} \quad (8.11)$$

and from (2.128) that the heat of reaction at constant volume is

$$Q_{VS} = \Delta E_o + ne\left(T_s\right)_{CO_2} + \frac{m}{2}e\left(T_s\right)_{H_2O} - e\left(T_s\right)_{C_nH_m} \quad (8.12)$$

where T_1, T_2, T_s are the temperatures at times t_1 and t_2, and the reference temperature for the lower heat of reaction Q_{VS}, respectively, then the first law of thermodynamics for the combustion powers in time step dt is

$$dQ - dW = (E(T_2) - E_2(T_s)) - (E(T_1) - E_1(T_s)) + dM_f\, Q_{VS}. \quad (8.13)$$

In this expression dQ is the heat transferred to or from the cylinder walls, dW is the work due to the piston motion, Q_{VS} is the lower heat of reaction and this is negative for an exothermic reaction, dM_f is the fuel burnt in the time step and the two bracketed terms represent the internal energies at states 2 and 1 measured above the reference temperature T_s. Expression (8.13) is completely general and can be used for the whole cycle, with dM_f zero when necessary, such as during the compression stroke.

In conventional compression ignition cycle calculations the lowest heat of reaction is replaced by the <u>lower calorific value</u> which is a <u>positive</u> number in an exothermic reaction.

In general engineering practice the term calorific value is used to denote the heating value of a fuel and is expressed in terms of conventional units of mass not in terms of mols. Thus the calorific value of a fuel would be expressed as a number of kJ/kg. Frequently the value used is the higher calorific value of the fuel. If so the lower calorific value must be estimated for use in these calculations.

In this exposition, a unit mol of fuel corresponds to 100† units of mass, hence the molar lower calorific value used here, q_{VS}, is 100 times the usual value available commercially, while the number of mols, dM_f, is one-hundredth of the number of units of mass burnt.

Then $$q_{VS} = -Q_{VS} \qquad (8.14)$$

and the first law is of the form

$$dQ-dW = (E(T_2)-E_2(T_s)) - (E(T_1)-E_1(T_s))-dM_f q_{VS}. \qquad (8.15)$$

The product $dM_f\ q_{VS}$ is called the "heat release". The term was devised to obtain a quantitative assessment of the combustion process. Various expressions have been developed as described in Chapter 4. The "heat release" is nominally expressed in the form of a mass rate equation, since q_{VS} is a constant. A number of expressions have been proposed, we shall use those developed at UMIST called the Whitehouse-Way equations. Another form of "heat release" expression is the "cumulative heat release pattern". This is simply the integrated fuel burnt, normally expressed as a fraction of the total fuel injected, against crankangle. Thus we can obtain dM_f either from a rate equation or an integrated equation; either way we can express dM_f as a function of the total fuel injected M_f in the form

$$dM_f = M_f dx, \qquad (8.16)$$

where $$x = x(\alpha).$$

In Section 8.4, when describing the "real" engine cycle calculation, we shall refer to the functional form of x. Before examining the real cycle we shall discuss an ideal cycle calculation with variable specific heats and composition with instantaneous combustion.

We shall develop the cycle calculation step by step with emphasis both on the thermodynamics and the numerical solution.

---

†The number of atoms of carbon n and hydrogen m are arranged so that the molecular weight of one mol of C_nH_m is 100 kg.

8.3 THE IDEAL DUAL-COMBUSTION CYCLE

The pressure volume diagram for an ideal cycle is shown in Fig. 8.1(a) and the corresponding internal energy temperature diagram in Fig. 8.1(b). The volume ratio of compression is V_A/V_B and the volume ratio of expansion V_E/V_D. In the ideal cycle fuel may be taken to be injected at two points only, namely at B and at C. The cycle of events are isentropic compression of air from A to B, adiabatic constant volume combustion from B to C, adiabatic constant pressure combustion from C to D and isentropic expansion from D to E. The cycle is closed at constant volume.

In the cycle calculations we shall use the first law of thermodynamics in the form given in expression (8.13) with Q_{VS} the heat of reaction at temperature T_s.

8.3.1 Isentropic Compression

A charge of pure air is compressed isentropically from volume A to volume B. To calculate the state changes p and T we subdivide the stroke volume into a number of intervals of small volume change and equate the correlations at the beginning and end of each interval using the first law of thermodynamics. The smaller the volume change in the interval the more accurate the calculation. We shall use the subscripts 1 and 2 to define the states at the beginning and the end of the volume change.

Consider a small change in volume $dV = V_2 - V_1$. We can apply the first law in the form given in (8.13), noting that for an isentropic process there is no heat transfer $dQ = 0$, and since there is no combustion and the composition of the air is not changed, then $dM_f = 0$ and $E_2(T_s) = E_1(T_s)$. Hence the first law is

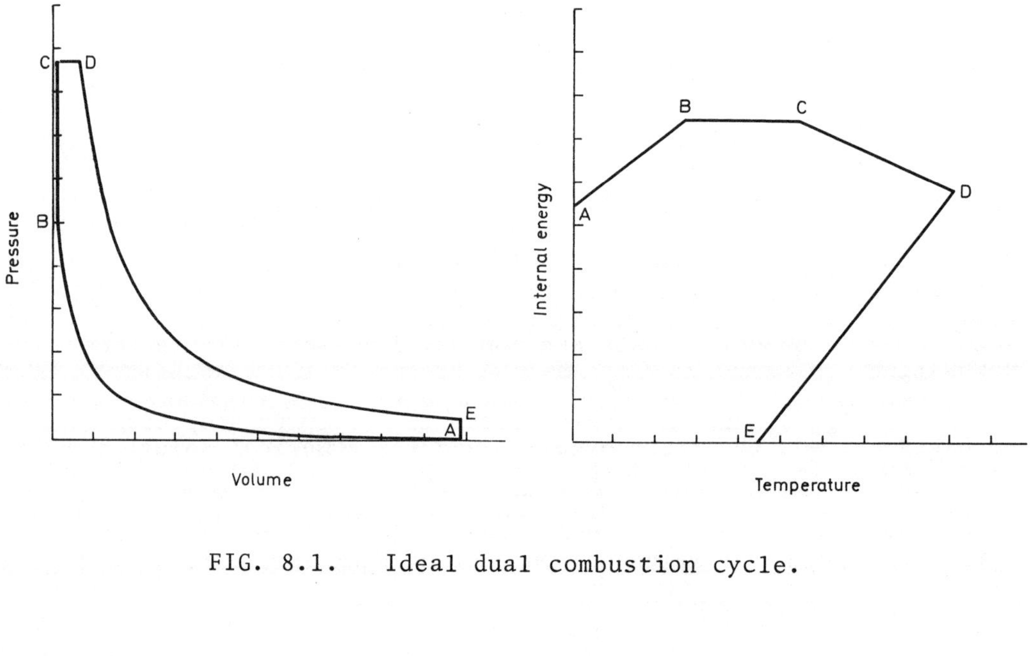

FIG. 8.1. Ideal dual combustion cycle.

$$-dW = E(T_2)-E(T_1). \tag{8.17}$$

The change in pressure will be from p_1 to p_2. If the volume change is small the work term can be given approximately by

$$dW = p\ dV = \left(\frac{p_1+p_2}{2}\right)(V_2-V_1). \tag{8.18}$$

In this expression we use the mean pressure during the volume change. We can substitute for dW into (8.17) and rearrange to give the expression

$$f(E) = E(T_2)-E(T_1) + \left(\frac{p_1+p_2}{2}\right)(V_2-V_1) = 0. \tag{8.19}$$

In this expression the initial conditions, subscript 1, will all be known. The unknowns will be the pressure p_2 and the temperature T_2. If T_2 is known $E(T_2)$ will be known. We therefore require a second equation; this is the state equation in the form

$$p_2 = \left(\frac{V_1}{V_2}\right)\left(\frac{T_2}{T_1}\right) p_1. \tag{8.20}$$

Equations (8.19) and (8.20) cannot be solved analytically and a numerical solution must be sought. In principle this requires the estimation of T_2 to obtain p_2 from (8.20) and $E(T_2)$ from (8.7). Both p_2 and $E(T_2)$ are then substituted into (8.19). A solution is obtained when f(E) is zero. The numerical method used may be the Newton-Raphson method. In this method if $(T_2)_{n-1}$ is the estimated value of T_2, then a better approximation $(T_2)_n$ will be given by

$$(T_2)_n = (T_2)_{n-1} - \frac{f(E)_{n-1}}{f'(E)_{n-1}}, \tag{8.21}$$

where $f'(E) = \frac{df(E)}{dT}.$

To obtain f'(E) we assume $dW/dT = 0$, which is a reasonable approximation because dW is not very sensitive to T_2, and

$$f'(E) = \frac{dE(T_2)}{dT}$$

since $\frac{dE(T_1)}{dT}$ is zero.

Now $(dE(T_2))/dT$ is the instantaneous specific heat at constant volume at temperature T_2 multiplied by the number of mols M in the mixture.

We can then write

$$f'(E) = MC_v(T_2)$$

and (8.21) now becomes

$$(T_2)_n = (T_2)_{n-1} - \frac{f(E)_{n-1}}{MC_v(T_2)_{n-1}}. \tag{8.22}$$

For a volume change dV the conditions at state 1 at the beginning of the step dV will be equal to the conditions at state 2 at the end of the previous volume step dV. The first estimate of T_2 may be obtained by assuming an isentropic change from the state conditions at T_1; thus

$$T_2 = T_1\left(\frac{V_1}{V_2}\right)^{k-1} = T\left(\frac{V_1}{V_2}\right)^{\frac{R_{mol}}{C_v(T_1)}}. \tag{8.23}$$

The internal energies $E(T_1)$, $E(T_2)$ and the specific heats $C_v(T_1)$, $C_v(T_2)$ are calculated from the gas composition and temperatures by the methods outlined in Chapter 2 and Section 8.2. A numerical technique is described in Appendix II.A.

As for an ideal cycle no products of combustion are left as residuals, the amount of the various gases forming the cylinder contents will depend on the required overall air/fuel ratio. If the stoichiometric air/fuel ratio is $(A/F)_{st}$ and the required overall air/fuel ratio is (A/F), then the number of mols of the cylinder constituents in (8.2) and (8.3), b_1, b_2, b_3, b_4, and b_5 will, at the beginning of the compression stroke A, be

$$b_1 = 0, \quad b_2 = 0, \quad b_3 = w\left(n + \left(\frac{m}{4}\right)\right)\frac{(A/F)}{(A/F)_{st}}$$

$$b_4 = 3.76 b_3, \quad b_5 = 0$$

and, since there is no combustion, $a_n = b_n$.

The total number of mols of the fuel C_nH_m, is w. This will be injected in two parts, $x_f w$ at constant volume and $(1-x_f)w$ during the constant pressure process.

An algorithm for solving (8.19) to (8.22) is given in Fig. 8.2.

The calculation proceeds in steps from V_A to V_B.

8.3.2 Adiabatic Combustion at Constant Volume

In the ideal cycle a portion of the fuel $x_f w$ is burnt in the constant volume period. Since the volume is constant no work is done and dW = 0. The first law for the combustion period is expression (8.13) in the form

$$0 = (E(T_2)-E_2(T_s))-(E(T_1)-E_2(T_s)) + x_f w\, Q_{vs}. \tag{8.24}$$

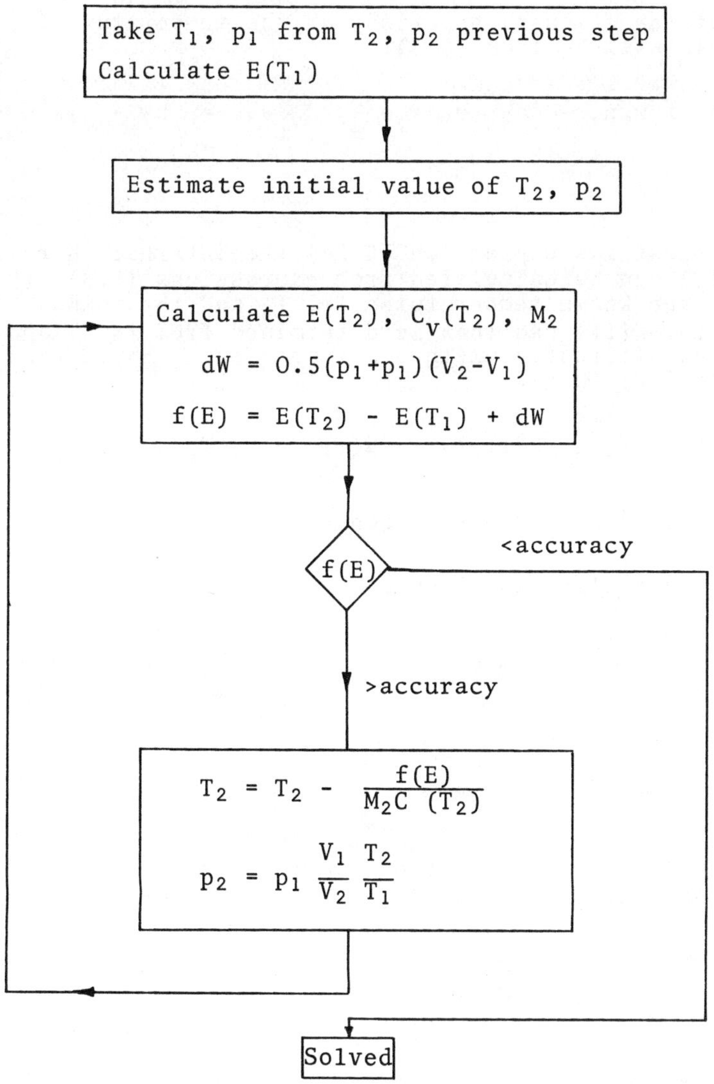

FIG. 8.2. Algorithm for compression and expansion strokes.

In this case the combustion is without heat loss and $dQ = 0$.

At the beginning of the combustion period the number of mols will be

$$b_1 = 0, \quad b_2 = 0, \quad b_3 = w\left(n + \frac{m}{4}\right) \frac{(A/F)}{(A/F)_{st}},$$

$$b_4 = 3.76\, b_3, \quad b_5 = x_f w.$$

At the end of the combustion period, at the temperature T_2, the number of mols will be from (8.3):

$$a_1 = x_f w\, n, \qquad a_2 = x_f w \frac{m}{2}, \qquad a_3 = b_3 - x_f w\left(n + \frac{m}{4}\right),$$

$$a_4 = b_4, \qquad a_5 = 0.$$

In the first law expression (8.24) the internal energies $E(T_1)$, $E_2(T_s)$, $E_1(T_s)$ can be calculated from expressions (8.8), (8.9) and (8.10) from the known temperatures T_1, T_s and the composition. The internal energy $E(T_2)$ can then be determined from (8.7) and hence T_2. In practice an alternative method may be used. Expression (8.24) is equated to f(E) to give

$$f(E) = (E(T_2) - E_2(T_s)) - (E(T_1) - E_2(T_s)) + x_f Q_{VS} = 0, \qquad (8.25)$$

and this is solved by Newton-Raphson's method.

In this case f'(E) is given by

$$f'(E) = \frac{dE(T_2)}{dT} = M_2 C_v(T_2) \qquad (8.26)$$

since

$$\frac{dE_2(T_s)}{dT} = \frac{dE_1(T_s)}{dT} = \frac{dE(T_1)}{dT} = 0$$

and $C_v(T_2)$ is the specific heat at constant volume at T_2 of the mixture of gases after combustion. The number of mols of gases M_1 and M_2 are given by

$$M_1 = \sum_{i=1}^{i=4} b_n, \qquad M_2 = \sum_{i=1}^{i=4} a_n. \qquad (8.27)$$

It will be noted that as the fuel is a liquid it does not contribute to the number of mols of gases.

The first estimate of the temperature T_2 may be given by

$$T_2 = T_1 - \frac{x_f w\, Q_{VS}}{M\, C_v(T_1)}. \qquad (8.28)$$

Again, $C_v(T_1)$ is the specific heat of gases after combustion at temperature T_1.

The new value of T_2 is then obtained from

$$(T_2)_n = (T_2)_{n-1} - \frac{f(E)}{M_2 C_v (T_2)_{n-1}} \tag{8.29}$$

The final pressure is given by

$$p_2 = \left(\frac{M_2}{M_1}\right)\left(\frac{T_2}{T_1}\right) p_1 . \tag{8.30}$$

The algorithm for this calculation is given in Fig. 8.3.

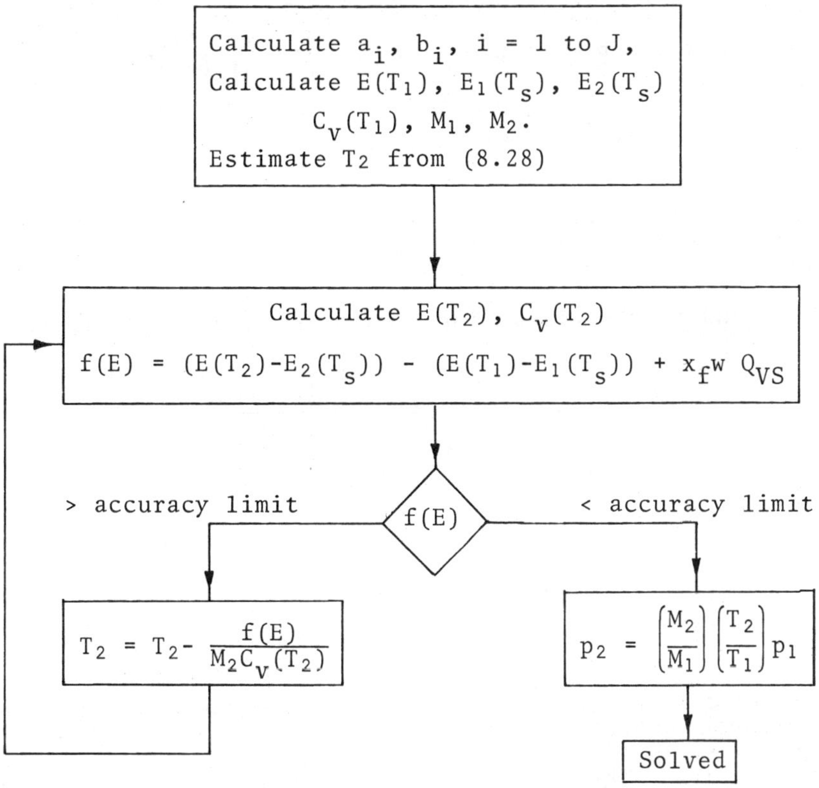

FIG. 8.3. Algorithm for constant volume combustion.

The temperature T_1 corresponds to the temperature at B and the temperature T_2 corresponds to the temperature at C. Thus only one step is required in the calculation for constant volume combustion.

8.3.3 Adiatabic Combustion at Constant Pressure

The remainder of the fuel, $(1-x_f)w$ mols, is injected and burnt from the end of the constant volume period (at C). Combustion is at constant pressure, so there is a volume increase from V_1 to V_2 with the resultant work, $dW = p(V_2-V_1)$. For adiabatic combustion, i.e. with no heat transfer, $dQ = 0$, and the first law, expression (8.13), is

$$-p_1(V_2-V_1) = (E(T_2)-E_2(T_s)) - (E(T_1)-E_1(T_s))+(1-x_f)w\, Q_{VS}. \qquad (8.31)$$

At the commencement of this period the temperature is T_1 and the number of mols from (8.2) will be:

$$b_1 = x_f wn, \quad b_2 = x_f w\frac{m}{2}, \quad b_3 = w\left(n+\frac{m}{4}\right)\left(\frac{(A/F)}{(A/F)_{st}} - x_f\right)$$

$$b_4 = 3.76w\left(n+\frac{m}{4}\right)\left(\frac{(A/F)_{st}}{(A/F)}\right), \quad b_5 = (1-x_f)w.$$

Here we notice that carbon dioxide (CO_2) and water vapour (H_2O are present in addition to oxygen (O_2) nitrogen (N_2) and fuel (C_nH_m) at the beginning of the combustion period.

At the end of the combustion period the number of mols in (8.3) will be

$$a_1 = wn, \quad a_2 = w\frac{m}{2}, \quad a_3 = w\left(n+\frac{m}{4}\right)\left(\frac{(A/F)}{(A/F)_{st}} - 1\right),$$

$$a_4 = b_4, \quad a_5 = 0.$$

The total number of mols of gases at 1 and 2 will be

$$M_1 = \sum_{i=1}^{i=4} b_i, \quad M_2 = \sum_{i=1}^{i=4} a_i. \qquad (8.32)$$

The temperature T_1 corresponds to the temperature at the end of the constant volume combustion, the internal energy terms $E(T_1)$, $E_1(T_s)$, $E_2(T_s)$ in expression (8.31) can therefore be evaluated from (8.8), (8.9) and (8.10). The unknowns in the first law are $E(T_2)$, i.e. T_2, and the volume V_2. The second equation for V_2 is from the state equation

$$V_2 = \left(\frac{M_2}{M_1}\right)\left(\frac{T_2}{T_1}\right) V \qquad (8.33)$$

for a constant pressure process.

The numerical solution is as before. The first law (8.31) is rearranged in the form

$$f(E) = (E(T_2)-E_2(T_s)) - (E(T_1)-E_1(T_s)) + (1-x_f)wQ_{VS} + p_1(V_2-V_1), \tag{8.34}$$

the derivative of (8.34) is

$$f'(E) = \frac{dE(T_2)}{dT} = M_2C_v(T_2) \tag{8.35}$$

and the Newton-Raphson expression

$$(T_2)_n = (T_2)_{n-1} - \frac{f(E)}{M_2C_v(T_2)}. \tag{8.36}$$

The first estimate for T_2 is given by

$$T_2 = T_1 - \frac{(1-x_f)wQ_{VS}}{M_2C_v(T_1)} \tag{8.37}$$

The algorithm for solving (8.34) and (8.36) is given in Fig. 8.4.

The temperature T_1 corresponds to T_C and the temperature T_2 to T_D. The calculation for the constant pressure period is carried out in one step.

8.3.4 Isentropic Expansion

During the expansion stroke from V_D to V_E the composition of the cylinder contents are constant. Thus the calculations of the state changes are exactly the same as those for the compression stroke with the number of mols for each constituent fixed by the conditions at the end of combustion, state D. The algorithm in Fig. 8.2 is therefore used for the expansion process.

The cycle closes at A by closing the pressure/volume diagram at constant volume (Fig. 8.1). The work done in the cycle is obtained by summing the work terms for each step, and is

$$W = \oint_{V}^{V_A} p\, dV \tag{8.38}$$

and the mean effective pressure is

$$p_m = \frac{W}{V_A-V_B}. \tag{8.39}$$

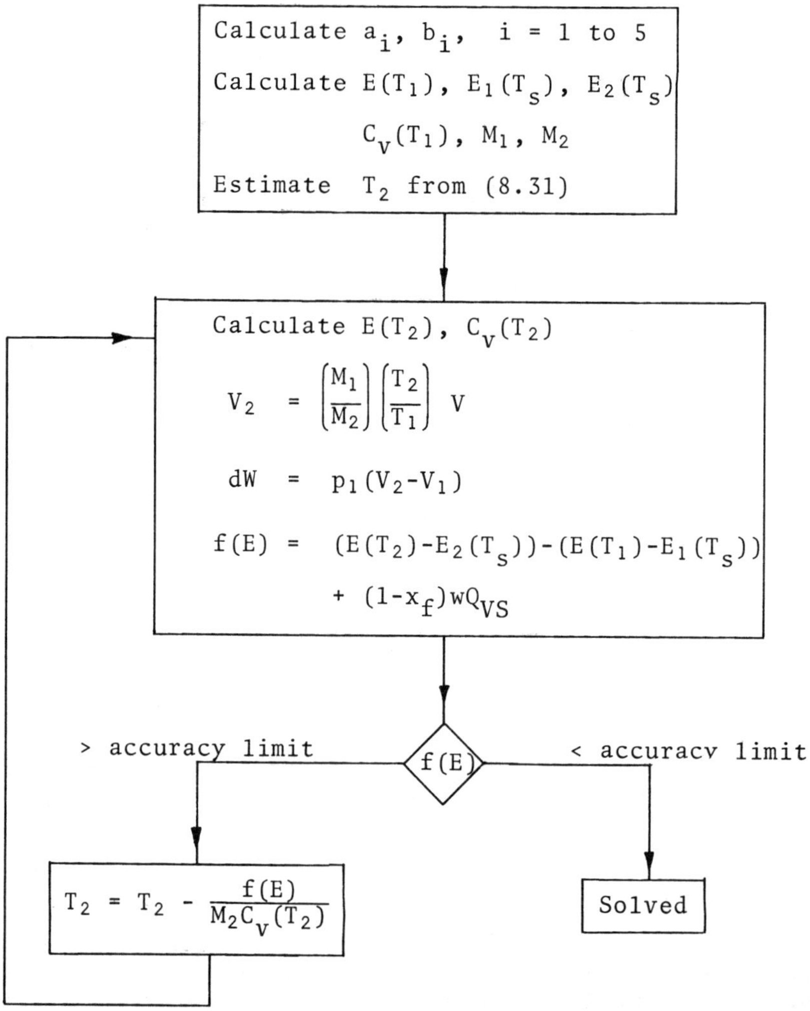

FIG. 8.4. Algorithm for constant pressure combustion

The thermal efficiency η_{TH} is

$$\eta_{TH} = -\frac{W}{wQ_{VS}} \times 100\%. \tag{8.40}$$

The minus sign allows for the sign of the heat of reaction Q_{VS}.

8.3.5 Cycle Studies

The methods outlined in the previous section may be integrated to provide the complete cycle. In Appendix II.B a FORTRAN listing is given for such a cycle. In Figs. 8.5 and 8.6 some results are shown of cycle studies using the program. You will see (Fig. 8.5) that as the air/fuel ratio decreases so more work is carried out in the cylinder (mean effective pressure rises) but that the energy transfer is less efficient (thermal efficiency decreases). The increase in indicated mean effective pressure is due to the increase in fuel supplied, the decrease in thermal efficiency is due to the increase in specific heat of the gas mixture with temperature and amount of combustion products and also to changes in the effective

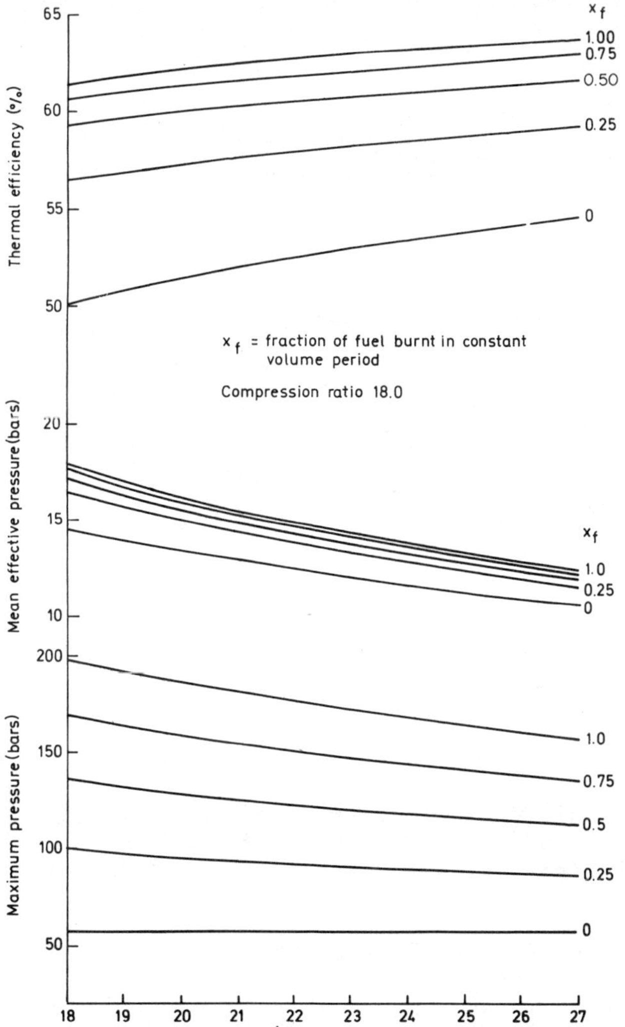

FIG. 8.5. Influence of fuel burnt and proportional constant volume period on indicated thermal efficiency, indicated mean effective pressure and maximum pressure.

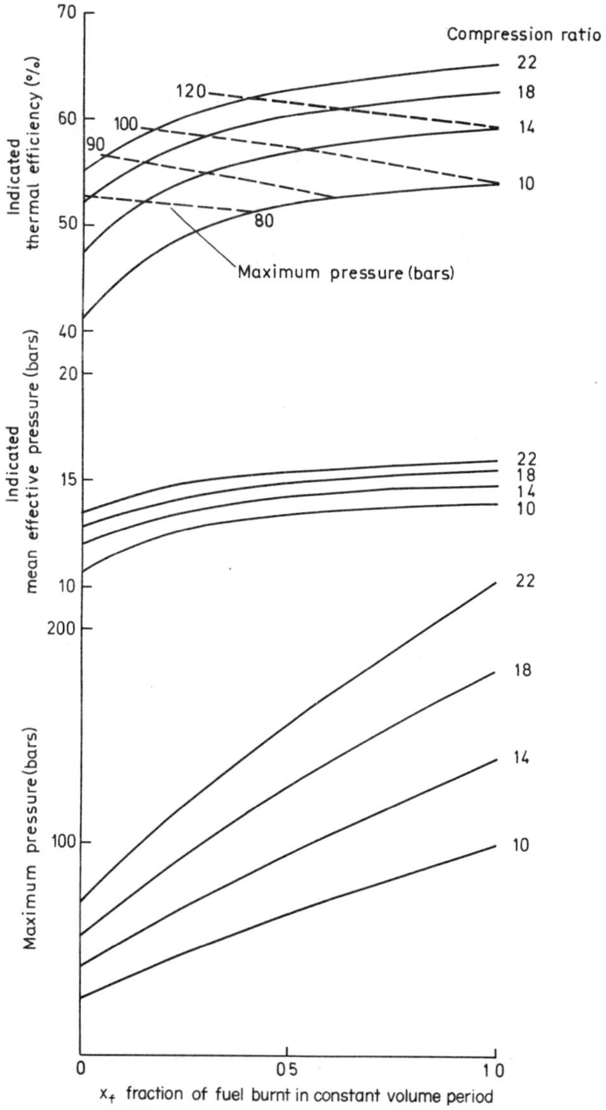

FIG. 8.6. Influence of compression ratio on indicated thermal efficiency indicated mean effective pressure and maximum cylinder pressure, for a constant air/fuel ratio (21:1).

expansion ratio of the constant pressure part of the cycle with amount of fuel burnt at constant pressure. The first effect dominates as x_f approaches unity (constant volume cycle) but the second is most important as x_f approaches zero (constant pressure cycle). The more combustion occurs at constant volume the better the indicated efficiency but the higher the maximum pressure. In Fig. 8.6 the influence of compression ratio is shown on the cycle thermal efficiency and work. Here we notice in the upper set of graphs, where lines of constant maximum pressure are shown, that for a given air/fuel ratio and maximum pressure the thermal efficiency <u>increases</u> with increase in constant pressure combustion ($x_f \to 0$). This is due to the influence of expansion ratio on the efficiency as discussed in Chapter 3 on air standard cycles. This is clearly illustrated in Fig. 8.7.

The ideal cycle does not take into account heat transfer and the variable fuel injection rate. To do so we must use more complex studies. In the next section we shall discuss cycle calculations which approximate more closely to real engine cycles. In this cycle we shall assume a single-zone combustion model.

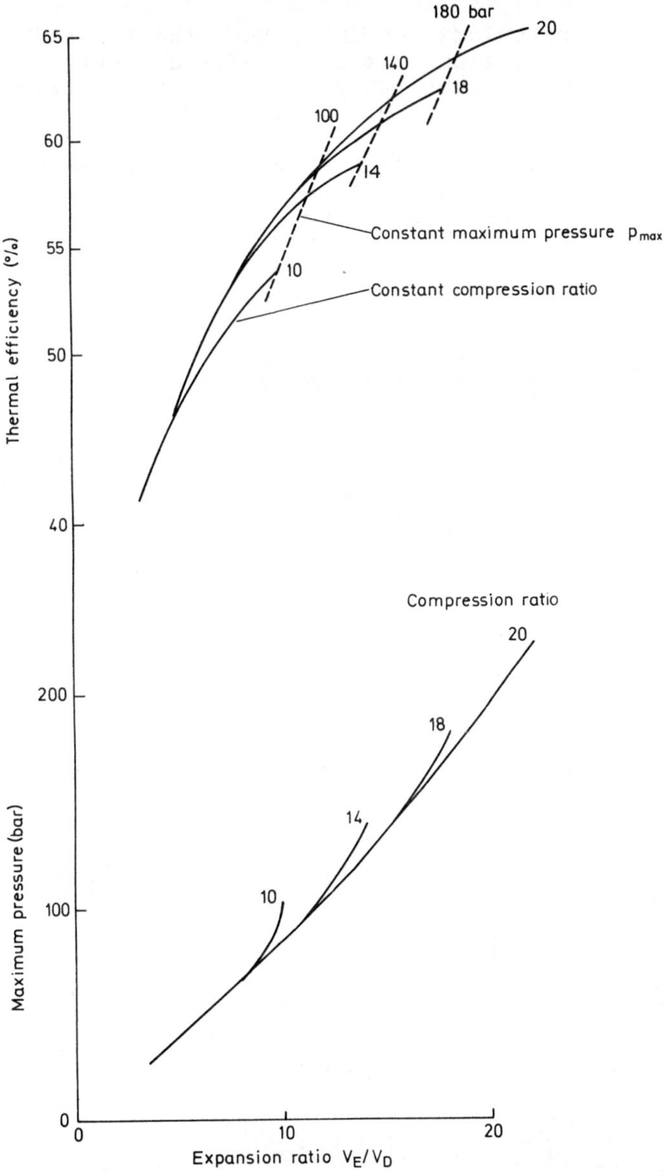

FIG. 8.7. Influence of expansion ratio on thermal efficiency.

8.4 REAL CYCLE WITH SINGLE-ZONE COMBUSTION MODEL

We will turn to the "real" cycle. Cycle calculations are subdivided into two parts—the closed cycle and the open cycle. The former comprises the period when both the inlet valves (or ports) and the exhaust valves (or ports) are closed and may be called the closed cylinder power process, and the latter is the period when either the inlet valves (or ports) or the exhaust valves (or ports) or both are open, this is called the gas exchange process. In this chapter we shall only be concerned with the closed cylinder power process.

To start the calculation we must know the trapped pressure, temperature and composition of the cylinder contents. These will be dependent on the gas exchange process so that, strictly speaking, a closed cycle calculation cannot be carried out in isolation.

The engine cycle calculation is by a step-by-step method, as in the ideal cycle, but in this case the variable is not the change in volume V but the crankangle α. The time step $d\alpha$ is usually one or two degrees, depending on the accuracy required. The cylinder volume V is related to the cylinder bore D, the stroke S, the connecting rod length ℓ and the clearance volume V_c by the expression

$$V = V_c + \left(\frac{\pi D^2}{4}\right)\left(\frac{S}{2}\right)(1 + n - (n^2 - \sin^2\alpha)^{\frac{1}{2}} - \cos\alpha), \qquad (8.41)$$

where n is the connecting rod length ℓ divided by the crank radius S/2.

The clearance volume V_c is related to the swept volume V_s through the volume ratio of compression, or simply the compression ratio CR, given by

$$CR = \frac{V_s + V_c}{V_c}. \qquad (8.42)$$

For each time step the first law of thermodynamics is set up in the form

$$dQ - dW = dE$$

$$\text{or} \quad dQ - dW = (E(T_2) - E_2(T_s)) - (E(T_1) - E_1(T_s)) - dM_f\, q_{VS}. \qquad (8.43)$$

Here we use the second form of the first law expression (8.15). The first term dQ represents the heat transfer to the cylinder contents from the cylinder walls. If the cylinder wall temperature is T_W and the exposed surface area is A_S, then, using Annand's equation (Chapter 6), the heat transfer is

$$\frac{dQ}{dt} = \left| ak\,\frac{(Re)^b}{D}\left(T_c - T_w\right) + c\left(T_c^4 - T_w^4\right)\right| \; J/s. \qquad (8.44)$$

In this expression a, b and c are constants and T_c is the mean temperature of the cylinder contents during a time step.

The Reynolds number Re is defined as

$$Re = \rho \frac{DV_p}{\mu}, \tag{8.45}$$

where V_p is the mean piston velocity given by

$$V_p = 2NS, \tag{8.46}$$

where N is the engine speed in revolutions per second. The viscosity μ will depend on the composition of the cylinder contents and the temperature.

The thermal conductivity k is related to the specific heat and viscosity for a Prandtl number of 0.7 by the expression

$$k = \frac{C_p\mu}{0.7}. \tag{8.47}$$

To convert (8.44) into the step value dQ we use the expression

$$dQ = \frac{dQ}{dt}\frac{dt}{d\alpha}\Delta\alpha = \frac{1}{360N}\frac{dQ}{dt}\Delta\alpha. \tag{8.48}$$

The work term dW is, as in the ideal cycle, evaluated from the mean cylinder pressure thus

$$dW = \left(\frac{p_1+p_2}{2}\right)\left(V_2-V_1\right). \tag{8.49}$$

For the combustion period the quantity of fuel burnt, dM_f, will depend on the heat release model used: we shall use two. The first, in the form of heat release "patterns", is shown in Fig. 8.8. Here the fuel quantity in a time step $\Delta\alpha$ is given by

either $dM_f = (x_2-x_1)w$

or $dM_f = dxw,$ (8.50)

where w is the total number mols of fuel injected per working stroke.

The "heat release" $dM_f\, wq_{VS}$ is then

$$(x_2-x_1)wq_{VS} \quad \text{or} \quad dx\, w\, q_{VS}. \tag{8.51}$$

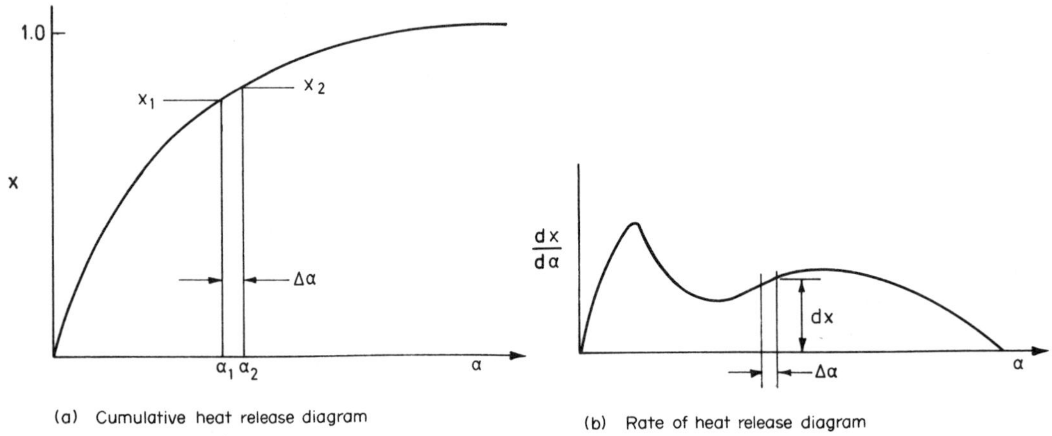

FIG. 8.8. Heat release diagrams.

The second method is based on the Whitehouse-Way model (Chapter 4). There are two expressions for fuel preparation and reaction rates.

The fuel preparation rate P is given by

$$P = KM_i^{1-x} M_u^x p_{O_2}. \quad \text{kg/deg,} \tag{8.52}$$

the reaction rate R by

$$R = \frac{K' p_O}{N\sqrt{T}} e^{\frac{-act}{T}} \int (P-R)\, d\alpha \quad \text{kg/deg.} \tag{8.53}$$

For the first expression (8.52) K, x, m are constants (see Table 4.4), M_i is the cumulative mass of fuel injected into the cylinder at any instant of time (or crankangle), M_u is the mass of unprepared fuel in the cylinder at that time, equal to $M_i - \int P\, d\alpha$, and p_{O_2} the partial pressure of oxygen in bars. In the second expression for the reaction rate, K' and act are constants (see Table 4.4), N is the engine speed ref/s, T the cylinder temperature, and $\int(P-R)d\alpha$ the summation of the difference between the fuel prepared and fuel burnt. In order to explain the application of the Whitehouse-Way equations to cycle calculations it is necessary to examine the model in some detail. Figure 8.9 shows the rate and cumulative diagrams corresponding to Fig. 8.8. In the cumulative diagrams, Fig. 8.9(b), we show the fuel injected, the fuel burnt and the fuel prepared.

The fuel injected is M_i.

The fuel prepared is $\int P\, d\alpha$.

The fuel burnt is $\int R\, d\alpha$.

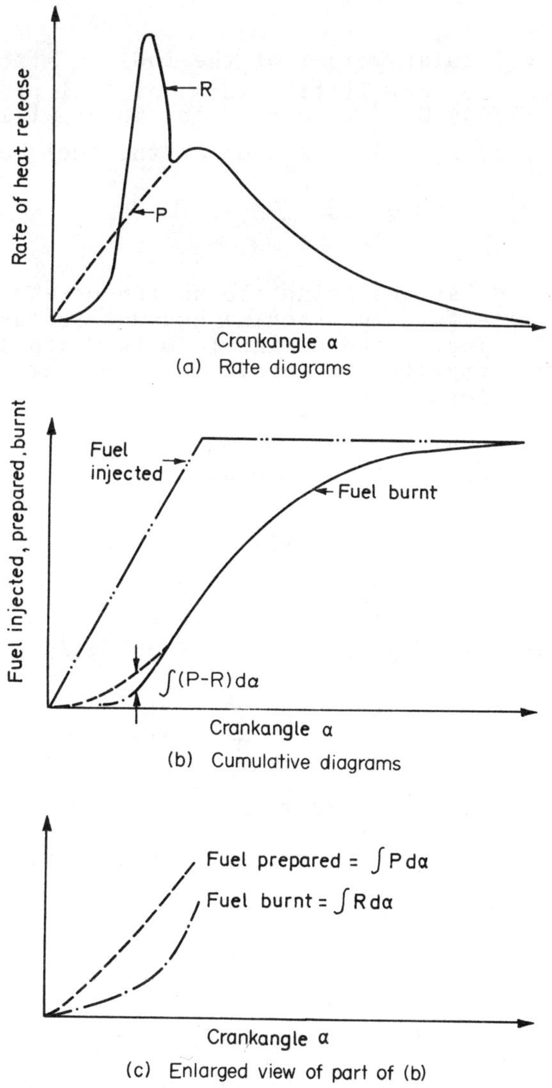

FIG. 8.9. Relation of heat release diagrams to fuel.

In the early part of the combustion period there will be an excess of fuel prepared over fuel burnt (see the enlarged view). This is given by

$$\int P \, d\alpha - \int R \, d\alpha = \int (P-R) d\alpha . \qquad (8.54)$$

During this period the fuel burnt in a step $\Delta\alpha$ is given by

$$dM_f = \frac{R\Delta\alpha}{W_m}, \tag{8.55}$$

where W_m is the molecular weight of the fuel. After a short period of time there will be very little excess of fuel prepared over fuel burnt, i.e. $\int(P-R)d\alpha \doteq 0$. Then the fuel burnt will be controlled by the preparation rate P, and in a step $\Delta\alpha$ the fuel burnt will be

$$dM_f = \frac{P\Delta\alpha}{W_m}. \tag{8.56}$$

Numerical calculations using the Whitehouse-Way models may be of two forms. A gross form in which no account is taken of separate elements of fuel injected and a fine form in which separate elements are examined and integrated to obtain the fuel burnt. We shall only describe the gross form.

If we consider the nth step of the calculation, the total fuel injected up to the beginning of the step is

$$\left(M_i\right)_{n-1} = \sum_1^{n-1} \frac{dM_i}{d\alpha}\Delta\alpha \tag{8.57a}$$

and the amount injected up to the end of the step will be

$$\left(M_i\right)_n = \sum_1^{n} \frac{dM_i}{d\alpha}\Delta\alpha, \tag{8.57b}$$

where $(dM_i/d\alpha)$ is the fuel injection rate, kg per degree.

Similarly, the total fuel prepared in the cylinder is given by

$$\left(P_T\right)_{n-1} = \sum_1^{n-1} P\Delta\alpha \tag{8.58a}$$

and

$$\left(P_T\right)_n = \sum_1^{n} P\Delta\alpha, \tag{8.58b}$$

and the total amount burnt up to the beginning of the step is

$$\left(R_T\right)_{n-1} = \sum_1^{n-1} R\Delta\alpha. \tag{8.59}$$

The mass of fuel prepared during this step is calculated from (8.51). In principle average values of M_i and M_u for the step would be the logical values to use, but this demands a knowledge of the conditions at the end of the step, which can only be obtained by an iterative procedure. If the change in the values during the step is small enough, relative to the absolute values, which is usually true for the step sizes used in these calculations, the initial values could be used.

For the first steps of the process, when M_i and M_u are very small, this condition is not valid and use of initial values alone would underestimate preparation. For this reason, as M_i is usually known and specified as data, the final value of M_i for the step may be used.

Then $M_i = (M_i)_n$ and $M_u = (M_i)_n - (P_T)_{n-1}$.

This gives the preparation rate during the step as

$$P_n = K\left(M_i\right)_n^{1-x}\left[(M_i)_n - (P_T)_{n-1}\right]^x p_{O_2} \quad \text{kg/deg.} \tag{8.60}$$

Then the total fuel prepared including that prepared in the time step is

$$(P_T)_n = (P_T)_{n-1} + P_n \Delta\alpha \text{ kg.} \tag{8.61}$$

To obtain the fuel burnt in the step we use the reaction rate equation (8.52). In this expression the $\int (P-R)\,d\alpha$ is $(P_T)_n - (R_T)_{n-1}$ and the reaction rate R_n for the step is

$$R_n = \frac{K' p_{O_2}}{N\sqrt{T}} e^{\frac{-act}{T}} \left[(P_T)_n - (R_T)_{n-1}\right] \text{ kg/deg.} \tag{8.62}$$

The notional fuel burnt including that burnt in the step under consideration is now

$$(R_T)_n = (R_T)_{n-1} + R_n \Delta\alpha \text{ kg.} \tag{8.63}$$

If $(R_T)_n$ is less than $(P_T)_n$ there is sufficient fuel prepared for (8.61) to hold and the fuel burnt in the step is

$$dM_f = R_n \Delta\alpha \text{ kg.} \tag{8.64}$$

If, however, $(R_T)_n$ is greater than $(P_T)_n$, there is insufficient prepared fuel in the cylinder for combustion to proceed in this way and combustion is assumed to be controlled by the rate of preparation. The fuel burnt is then

$$dM_f = P_n \Delta\alpha. \tag{8.65}$$

Once in a calculation the condition $(R_T)_n$ is greater than $(P_T)_n$ has been reached all subsequent combustion is preparation rate controlled (8.52) and the reaction rate expression is not used.

It might be argued that (8.62) by using the difference between the amount of fuel prepared at the end of the step and that burnt at the beginning of the step increases the reaction rate R_n unduly when this difference is very small, as it is at this changeover point at the end of pre-mixed burning. However, if the more valid mean values are used, i.e.

$$R_n = \frac{K' p_{O_2}}{N\sqrt{T}} e^{\frac{-act}{T}} \left(\frac{P_{T_n} + P_{T_{n-1}} - R_{T_{n-1}} - R_{T_n}}{2} \right),$$

the value of R_{T_n} is found to oscillate about the value of P_{T_n} so greatly during the necessary iterations that the calculation becomes unstable as the ( ) becoming alternately positive and negative.

As discussed earlier with respect to M_i and M_u, the values of T and p_{O_2} used in these rate equations should ideally be average values. As in some examples of multi-zone models of combustion p_{O_2} may be very small, and as the reaction rate is very temperature sensitive, it is usual to obtain these average values using iteration.

We are now in a position to describe a calculation procedure for a closed cycle. There is a number of ways for setting up a calculation; the method we propose to illustrate is similar to the ideal cycle calculation. A flow diagram is shown in Fig. 8.10. The calculation starts with the trapped conditions, pressure p_1, temperature T_1 and the composition of cylinder constituents b_i, which will depend on the air quantity supplied and the residuals from the products of combustion from the previous cycle. We set $a_i = b_i$ and with the known temperature T, evaluate the internal energy terms $E(T_1)$, $E_1(T_s)$ and $C_v(T_1)$ from (8.8) and (8.10).

From the angle at the commencement of the compression stroke we can determine the total cylinder volume V_1 from (8.41). The angle is then increased by $\Delta\alpha$ to give $\alpha_2 = \alpha_1 + \Delta\alpha$ and the volume V_2 at the end of the step determined. In the general calculation, to minimize the computing scheme, this angle is compared with the angle for nominal fuel injection α_I, and if it is greater than α_I the fuel burnt dM_f is evaluated. For the start of the calculation the angle α_2 will be less than α_I and so dM_f is set zero for the compression stroke. We now have all the known data and the next step is to solve the first law of thermodynamics (expression (8.43)) for the step. We first estimate the temperature at the end of the step T_2 from an approximate expression for the first law. This gives T_2 as

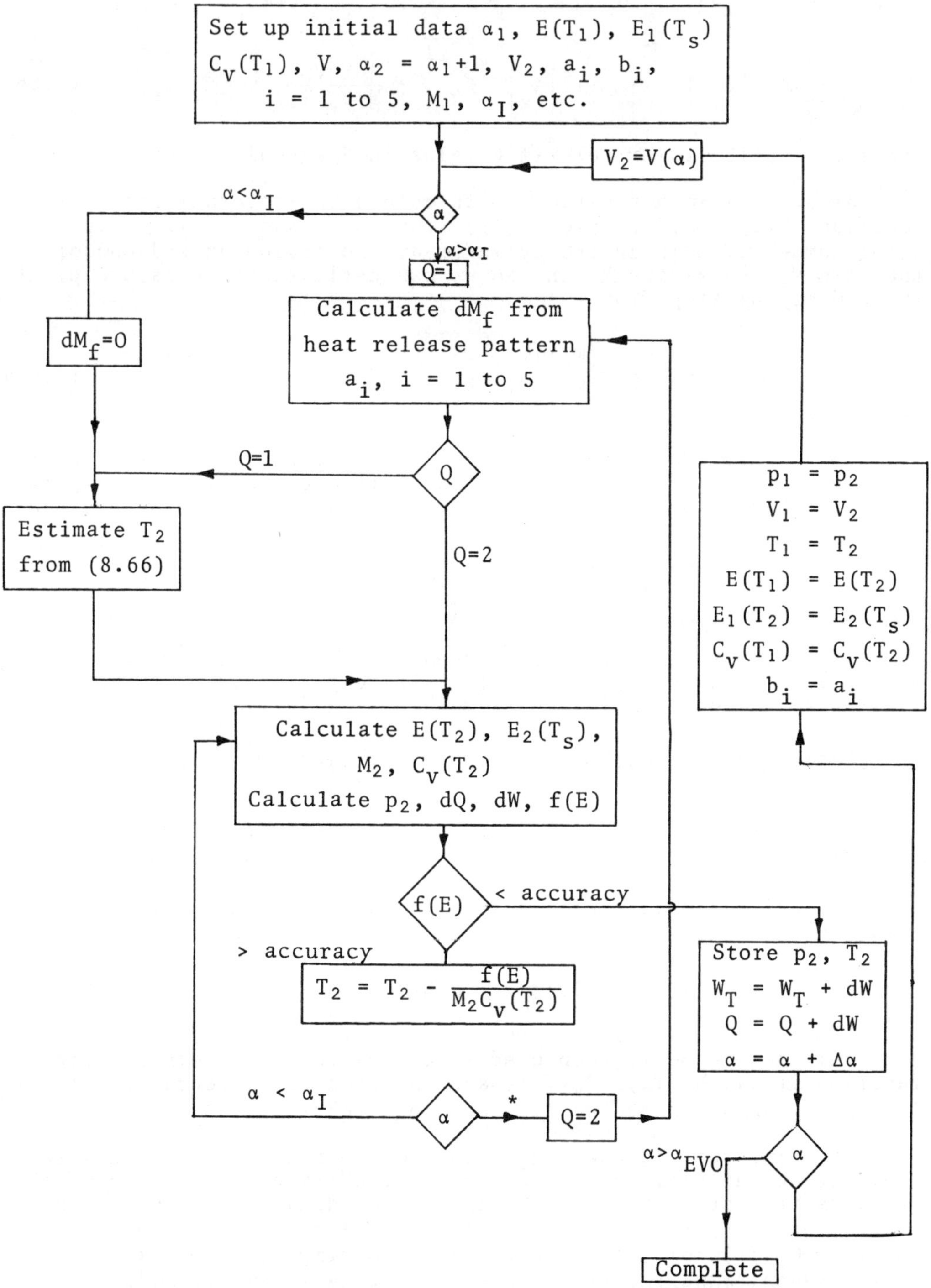

FIG. 8.10. Flow diagram for closed cycle calculation. * This test for Whitehouse-Way Model only, α_I = nominal injection angle, α_{EVO} = angle at exhaust valve open.

$$T_2 = T_1 \left(\frac{V_1}{V_2}\right)^{\frac{R_{mol}}{C_v(T_1)}} + \frac{dM_f q_{VS}}{M_1 C_v(T_1)}, \qquad (8.66)$$

where q_{VS} is the lower calorific value in J/kg-mol.

We are now in a position to calculate the internal energy functions $E(T_2)$, $E_2(T_s)$ and $C_v(T_2)$ from (8.7) and (8.9). The total number of mols in the cylinder at the beginning and end of the step M_1 and M_2 are known, so we can estimate the pressure p_2 at the end of the step from

$$p_2 = \left(\frac{M_2}{M_1}\right)\left(\frac{T_2}{T_1}\right) p_1. \qquad (8.67)$$

We now calculate the work done in the time step dW from (8.48) and the heat transfer from the walls dQ from expressions (8.44) to (8.47) with the mean cylinder temperature T_c during the step given by

$$T_c = \frac{T_1 + T_2}{2}.$$

If the correct value of T_2 is estimated, both sides of expression (8.43) should be equal. To test if this is so we rearrange the equation in the form

$$f(E) = (E(T_2)-E_2(T_s)) - (E(T_1)-E_1(T_s))+dW - dM_f q_{VS} - dQ \qquad (8.68)$$

If the numerical value of f(E) is less than the accuracy required, then the correct value of T_2 has been estimated, otherwise a new value of T_2 called $(T_2)_n$ is required; this is obtained from the original estimate $(T_2)_{n-1}$ using the Newton-Raphson method, namely

$$(T_2)_n = (T_2)_{n-1} - \frac{f(E)}{M_2 C_v(T_2)}. \qquad (8.69)$$

This new value is then used to calculate the internal energy function $E(T_2)$, $E_2(T_s)$, $C_v(T_2)$, and the process is repeated until T_2 is finally obtained to the accuracy required.

The time step is now completed and the next step is examined by stepping up the angle by $\Delta\alpha$ deg. The conditions at the end of the previous step are set up as the initial conditions for the new step. Thus $p_1 = p_2$, $V_1 = V_2$, $T_1 = T_2$, $E(T_1) = E(T_2)$, $E_1(T_s) = E_2(T_s)$, $C_v(T_1) = C_v(T_2)$, $b_i = a_i$. Where the subscript 1 refers to the initial conditions for the new step and subscript 2 the final conditions from the previous step.

The calculation is completed when the angle α is equal to or greater than the angle at which the exhaust valve (or ports) α_{EVC}. In a typical computer program the values of p and T are stored for each step and the total work and heat transfer calculated by successive summation of dW and dQ.

When the angle α exceeds the nominal injection angle α_I the combustion commences and the fuel burnt (dM_f) is calculated by one or other of the models described earlier. If the Whitehouse-Way model is used, the temperature in the rate equation corresponds to the mean cylinder temperature T_c. The iterative procedure in this case must include the calculation of dM_f as shown by the asterisked test in the flow diagram (Fig. 8.10).

During combustion the composition of the cylinder contents will change. Thus if the original composition at state 1 is

$$b_1CO + b_2H_2O + b_3O_2 + b_4N_2 + b_5C_nH_m,$$

then at state 2 the composition will be

$$a_1CO + a_2H_2O + a_3O_2 + a_4N_2 + a_5C_nH_m$$

and the values of these molal quantities will be

$$a_1 = b_1 + dM_f n, \qquad a_4 = b_4,$$

$$a_2 = b_2 + dM_f \frac{m}{2}, \qquad a_5 = 0,$$

$$a_3 = b_3 - dM_f \left(n + \frac{m}{4}\right), \qquad b_5 = dM_f.$$

It may be noted that, as before $M = \sum_1^4 b_i$, and the fuel is treated as a liquid not affecting other conditions until it is burnt. This is the most simple concept but one that does not allow for the cooling effect of the excess fuel injected, but not yet burnt, or especially as its temperature is raised for its true volume and/or partial pressure.

In Fig. 8.11 some typical pressure/crankangle diagrams are shown with various injection times. This clearly shows the influence of fuel injection timing on the cylinder diagram. The overall effect on power and efficiency is shown in Fig. 8.12. These results were obtained by a computer calculation using the Whitehouse-Way model for combustion.

This comparatively simple, single-zone model calculation has been found to give satisfactory performance calculations for many engines within limited ranges of operation, no doubt because of the semi-empirical coefficients used. In reality the cylinder contents are not homogeneous during combustion. To approach reality more

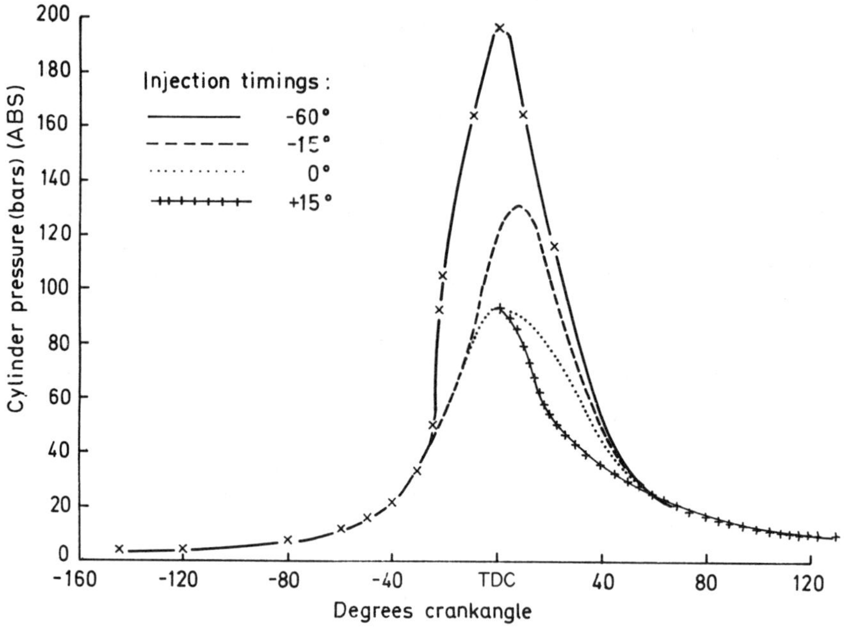

FIG. 8.11. Typical effect of injection timing on calculated cylinder pressure diagrams.

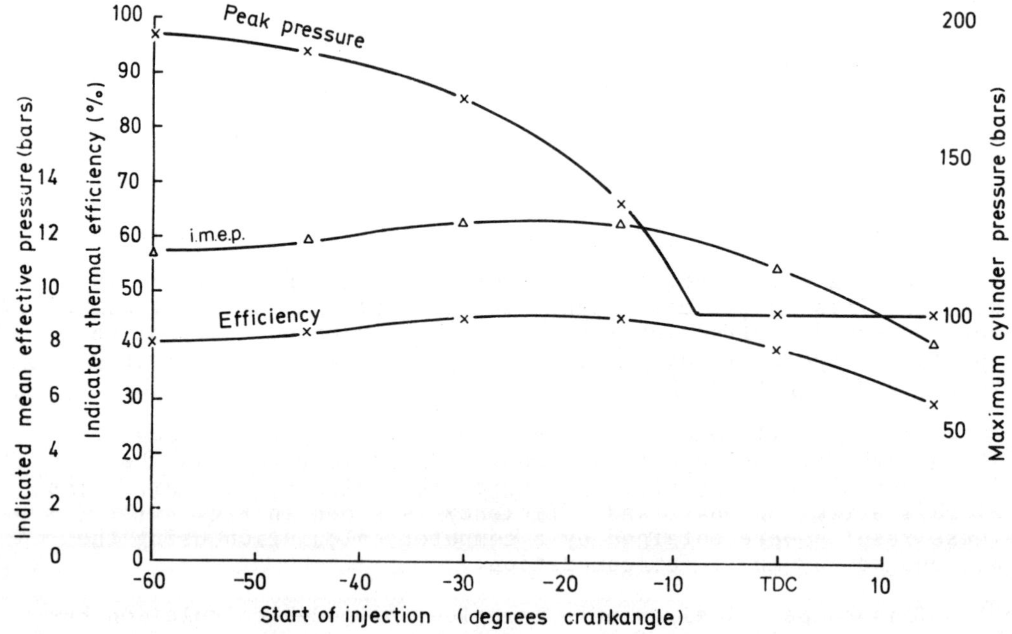

FIG. 8.12. Typical effect of injection timing on calculated engine performance.

nearly needs more sophisticated methods, such as multi-zone models for thermodynamic calculations with estimates of zonal inter-mixing (often referred to as air or gas entrainment). No conclusive method has yet been developed to deal with all of these complexities which are very dependent upon engine design. The next section deals with some of these matters.

8.5 MULTI-ZONE MODELLING

8.5.1 Thermodynamics of Two-zone Models

If we consider the general case of an engine cylinder with the contents divided up into two zones, each of which is itself homogeneous at any given time, each of these zones may be treated in a way basically similar to the single zone described in Section 8.4 with the addition of matter transfers between the zones and the consequent energy transfers.

Thus for two zones designated by subscripts A and B, the first law equation, corresponding to (8.43) in the single-zone case, we have

$$dQ_A - dW_A = (E(T_2) - E_2(T_s))_A - (E(T_1) - E_1(T_s))_A - dM_{fA}q_{VS} - dM_B h_B + dM_A h_A. \tag{8.70}$$

That is, the final internal energy is increased by the enthalpy of the mass transferred from the other zone and decreased by the enthalpy of the mass transferred to the other zone. There are consequent changes to the final mol count and structure in the first term.

Zone B has a similar, mirror image equation,

$$dQ_B - dW_B = (E(T_2) - E_2(T_s))_B - (E(T_1) - E_1(T_s))_B - dM_{fB}q_{VS} - dM_A h_A + dM_B h_B. \tag{8.71}$$

In principle, mass and enthalpy exchanges should be mean values, but initial values may be used with smaller step sizes for the same accuracy with consequent simplification of programming.

The work terms will be similar to the single-zone model except that the volume changes are those of the appropriate zone. So

$$dW_A = \left(\frac{p_1+p_2}{2}\right)\left(V_{2A}-V_{1A}\right), \qquad dW_B = \left(\frac{p_1+p_2}{2}\right)\left(V_{2B}-V_{1B}\right), \tag{8.72}$$

and as the total volume is that determined by the geometry of the engine, the total cylinder volume, we have

$$V_{2A}+V_{2B} = V_2, \qquad V_{1A}+V_{1B} = V_1.$$

The sum of the work done (dW_A+dW_B) equals the net work done on the piston dW required for the final estimate of engine work output.

In this case we have an extra unknown, for each zone, which was not an unknown for the single-zone model, that is the final volume of each zone. This may be dealt with by using the estimate of final pressure p_2 and the consequent energy balances to lead to the zone volumes V_{2A} and V_{2B} and then re-estimating the final pressure to correct any error in total final volume V_2 by the equation

$$p_2 = p_2 \frac{V_2}{V_{2A}+V_{2B}} \tag{8.73}$$

This gives a revised estimate of p_2 for the next iteration.

Existing two-zone models, as described in Chapter 4, simplify this general picture somewhat by having mass transfers only in one direction, from the air zone to the burning zone, and combustion occurs only in the burning zone. There is also a need to estimate heat losses from each zone to the walls and heat transfer between zones. These are uncertain matters and so far have been dealt with somewhat arbitrarily.

8.5.2 Multi-zone Models

The additional complexity introduced when more than two zones are involved is usually only considered if soot or gaseous pollutant formation is of interest. The simple complete combustion chemistry and the use of lower calorific value (or heat release) as in the single- and two-zone models described, is then unrealistic. It is better then to use total internal energies and total enthalpies of the various constituents (i.e. heats of formation at absolute zero plus enthalpy/internal energy above absolute zero). Thus the techniques for each zone are as described for the similar spark ignition case in Chapter 9, Section 9.2, which deals with rate kinetics. The mass and energy transfers are then similar in nature, though greater in number and complexity to those described in Section 8.5.1 above, and the chemical processes and complexities in each, or in practice usually only in some of the zones, may be as in Section 9.2.

Multi-zone calculations are as yet comparatively undeveloped because of the uncertainty about mixing rates and the problem of determining a compromise between the number of zones to use and the consequent cost of computing. Mention of some work in this area has been made in Chapter 4. The problem is thus more complicated than the spark ignition case mentioned in Chapter 9. There is no convenient flame front separating two homogeneous zones but a multiplicity of conditions consequent on the heterogeneity of the diesel combustion process. This complexity has to be arbitrarily simplified into a finite number of zones. Heat transfer from different zones is also an unknown, and at present dealt with in somewhat arbitrary ways. No doubt the literature will increase rapidly in the near future.

Chapter 9

Spark Ignition Engine Cycle Calculations

Notation

a_i	number of mols of species i per mol of fuel
C_p	specific heat at constant pressure
C_v	specific heat at constant volume
e	specific internal energy
E	internal energy
ff	flame factor
g	specific Gibbs function
h	specific enthalpy
H	enthalpy
k_{bi}	backward reaction rate constant for reaction i
k_{fi}	forward reaction rate constant for reaction i
K_p	equilibrium constant
m	number of hydrogen atoms in fuel, C_nH_m
m	mass of gas (Section 9.2)
M	number of mols
N	engine speed (rev/s)
n	number of carbon atoms in fuel, C_nH_m
p	pressure
P	normalized pressure = $\left(\frac{p}{\text{reference pressure}}\right) = \frac{p}{p_0}$
Q	heat transfer
R_i	one-way equilibrium rate for reaction i
R_{mol}	universal gas content
T	temperature
u_ℓ	laminar flame speed
u_t	turbulent flame speed
V	volume
w	number of mols of fuel
W	work
w_i	number of mols of species i
κ	specific heat ratio
α	crankangle or ratio of actual concentration of NO to equilibrium combustion
ϕ	equivalence ratio (= actual fuel:air ratio/ stoichiometric fuel:air fuel).
ν	stoichiometric coefficient

Suffixes

b	burnt zone
u	unburnt zone
R	reactant
P	product
0	level at zero temperature
1	calculation step 1
2	calculation step 2

The spark ignition engine works on the Otto cycle. In an ideal cycle a mixture of fuel and air is compressed, ignited and the combustion takes place at constant volume. The products of combustion expand and at the end of the expansion stroke the exhaust is at constant volume. In practice this ideal cycle is modified due to (a) heat losses in the compression, combustion and expansion periods, and (b) the finite time for the combustion processes. In the first part of the chapter we shall examine the ideal cycle and in the second part the real cycle. We shall confine the studies to hydrocarbon-air cycles. Since we have described the combustion processes in some detail in Chapter 5, the analyses will start directly with the chemical reactions and the corresponding thermodynamic properties of the substances using the expressions reviewed in Chapter 2.

Because the constituents in the cylinder during the combustion and the expansion stroke are dependent on the pressure and temperature, the thermodynamics of the ideal Otto cycle is more complex than the ideal dual combustion cycle discussed in Chapter 8. For this reason the combustion reactions are dealt with in more detail in this chapter than in the previous chapter.

9.1 IDEAL OTTO CYCLE WITH HYDROCARBON-AIR MIXTURE

In Chapter 3 we showed that the sequence of events for the ideal constant volume cycle with a gas with constant composition and constant specific heats were: (a) isentropic compression, (b) constant volume heat reception, (c) isentropic expansion and (d) constant volume heat rejection. The ideal Otto cycle with hydrocarbon-air mixtures differs from the ideal cycle (constant volume cycle) in a number of respects. The most important are the variation in gas composition during combustion and expansion and the variation of specific heats with temperature and composition, since in order to satisfy thermodynamic equilibrium dissociation takes place in the combustion process and reassociation in the expansion process.

In Fig. 9.1(a) the pressure/volume diagram is shown for an Otto cycle and in Fig. 9.1(b) the corresponding internal energy/ temperature diagram. The fuel/air mixture is compressed adiabatically from A to B. At B the mixture is ignited and combustion takes place from BC. This is an adiabatic constant volume process. The gas then expands adiabatically from C to D. The process from D to A is at constant volume. Due to the change in composition of the gas from A to D one cannot thermodynamically close the cycle DA since there will be a difference in entropy at D from the entropy at A and the thermodynamic properties will not be the same. We will simply "draw" a constant volume line for dA. The gas mixture in the cylinder is constant from A to B but of variable composition from B to D.

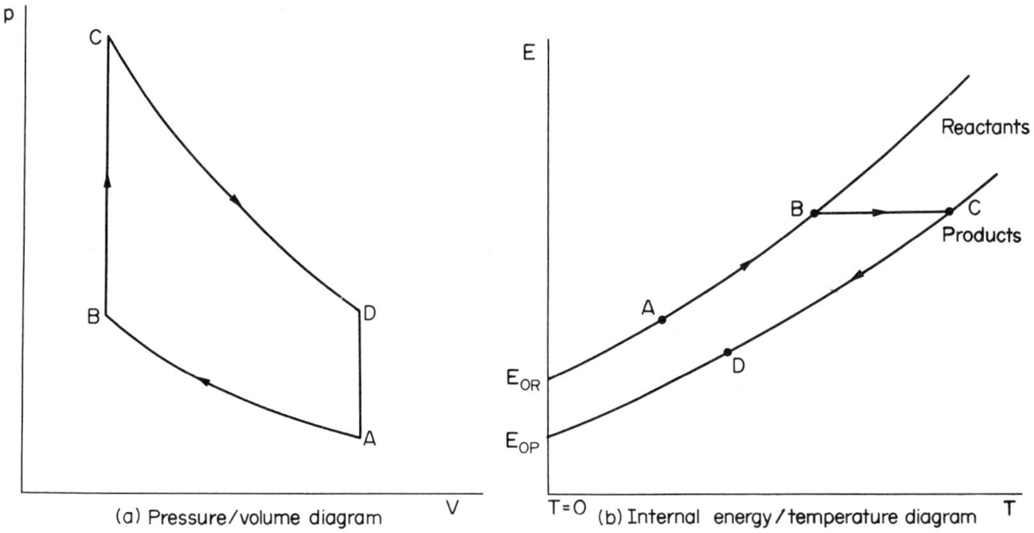

FIG. 9.1.

Normal fuels in spark ignition engines are mixtures of hydrocarbons. We can represent these fuels by a general hydrocarbon C_nH_m, where n is the number of atoms of carbon and m is the number of atoms of hydrogen. Conventional hydrocarbons fall into the following categories:

Paraffins	C_nH_{2n+2}	m = 2n + n
Naphthenes	C_nH_{2n}	m = 2n
Aromatics	C_nH_{2n-6}	m = 2n - 6

The well-known paraffins are:

Methane	n = 1	CH_4	Hexane	n = 6	C_6H_{14}
Ethane	n = 2	C_2H_6	Heptane	n = 7	C_7H_{16}
Propane	n = 3	C_3H_8	Octane	n = 8	C_8H_{18}
Butane	n = 4	C_4H_{10}			

Methane is a gas; propane and butane are called low pressure gases because under pressure they exist in liquid form whilst at atmospheric pressure they are gaseous. The higher molecular weight, paraffins, hexane, etc., are liquid at normal ambient temperature and pressure.

Well-known naphthenes are:

Cyclohexane	n = 6	C_6H_{12}
Hexahydrotoluene	n = 7	C_7H_{14}
Hexahydroxylene	n = 8	C_8H_{16}

Well-known aromatics are:

Benzenes	n = 6	C_6H_6
Toluene	n = 7	C_7H_8
Xylene	n = 8	C_8H_{10}

The basic stoichiometric chemical equation for a hydrocarbon-oxygen reaction is

$$C_nH_m + \left(n + \frac{m}{4}\right)O_2 = nCO_2 + \frac{m}{4}H_2O.$$

Now in air the ratio of the mols of O_2 to the mols of N_2 is given by

$$O_2 + \frac{79}{21}N_2 = O_2 + 3.76\,N_2.$$

Hence the basic stoichiometric equation for a hydrocarbon-air reaction is

$$C_nH_m + \left(n + \frac{m}{4}\right)O_2 + 3.76\left(n + \frac{m}{4}\right)N_2 = nCO_2 + \frac{m}{2}H_2O + 3.76\left(n + \frac{m}{4}\right)N_2.$$

The stoichiometric equation defines the <u>correct</u> mixture. To allow for mixtures different from the correct mixture we introduce the <u>equivalence ϕ</u>. This is defined as the ratio of the actual fuel/air ratio to the stoichiometric fuel/air ratio.

The stoichiometric fuel/air ratio is

$$\frac{C_nH_m}{\left(n + \frac{m}{4}\right)O_2 + 3.76\left(n + \frac{m}{4}\right)N_2}.$$

Hence for a mixture of equivalence ϕ, we have

$$C_nH_m + \left(n + \frac{m}{4}\right)\frac{1}{\phi}O_2 + 3.76\left(n + \frac{m}{4}\right)\frac{1}{\phi}N_2.$$

If the equivalence ϕ is <u>greater</u> than unity, the mixture is said to be <u>rich</u>, and if ϕ is less than unity the mixture is said to be <u>weak</u>. (Notice spark ignition engines may normally run with both rich <u>and</u> weak mixtures; on the other hand, compression ignition engines normally run with weak mixtures only.)

We are now in a position to define the hydrocarbon-air mixture at point A in the cycle (Fig. 9.1). For w mols of C_nH_m the mixture will be

$$w\left(C_nH_m + \left(n + \frac{m}{4}\right)\frac{1}{\phi}\,O_2 + 3.76\left(n + \frac{m}{4}\right)\frac{1}{\phi}\,N_2\right),$$

and the total number of mols M_A of the mixture is

$$M_A = w\left(1 + 4.76\left(n + \frac{m}{4}\right)\frac{1}{\phi}\right). \tag{9.1}$$

The magnitude of w is obtained from the ideal gas equation (2.1) for the mixture

$$p_A V_A = M_A\, R_{mol} T_A. \tag{9.2}$$

The cycle calculation is simply carried out by setting up the first law of thermodynamics for each step and solving for the change in state. For the combustion and expansion stroke a second set of equations is required to represent thermal equilibrium for the chemical reaction; these are the dissociation equations.

The same procedure is used for the cycle calculation as for the ideal compression ignition cycle in Chapter 8. The numerical solutions are based on the Newton-Raphson method. During the combustion process, i.e. from B to C, the calculations are extremely complex. An outline of the solution is given; for full details the reader is referred to reference (1).

We shall now examine the cycle step by step.

9.1.1 Adiabatic Compression

This corresponds to the period from A to B in the pressure and temperature diagrams (Fig. 9.1).

Consider a small change in volume from V_1 to V_2. The first law for the change (2.6) is

$$dQ - dW = dE,$$

and since the process is adiabatic, $dQ = 0$, we have

$$dE + dW = 0. \tag{9.3}$$

For a change in pressure from p_1 to p_2 the work dW is approximately

$$dW = p\,dV \doteqdot \left(\frac{p_1 + p_2}{2}\right)\left(V_2 - V_1\right). \tag{9.4}$$

The internal energy change is dE, which is given by

$$dE = E_2 - E_1. \tag{9.5}$$

The internal energy of the mixture is a function of the composition of the cylinder contents and temperature. Unlike a compression ignition engine the cylinder contents contain both fuel and air during the compression stroke. The internal energy is therefore the internal energy of the reactants in the hydrocarbon-air combustion process. The composition is considered to be constant during the compression stroke.

If E_R is the internal energy of the reactants, then from (2.106) we can write

$$E_R = E_{OR} + E_R(T).$$

If $\left(e_0\right)_{C_nH_m}$, $\left(e_0\right)_{O_2}$, $\left(e_0\right)_{N_2}$ are the specific internal energies at absolute zero for C_nH_m, O_2 and N_2, then the internal energy for the reactants at absolute zero E_{OR} is

$$E_{OR} = w\left(\left(e_0\right)_{C_nH_m} + \left(n + \frac{m}{4}\right)\frac{1}{\phi}\left(e_0\right)_{O_2} + 3.76\left(n + \frac{m}{4}\right)\frac{1}{\phi}\left(e_0\right)_{N_2}\right).$$

From (2.102) $h_0 = e_0$ and E_{OR} can be calculated from the coefficient $u_{i,7}$ in Table 2.1.

If the specific internal energies, relative to absolute zero, e(T) are $e(T)_{C_nH_m}$, $e(T)_{O_2}$ and $e(T)_{N_2}$, for C_nH_m, O_2 and N_2, then $E_R(T)$ for the mixture is given by

$$E_R(T) = w\left(e(T)_{C_nH_m} + \left(n + \frac{m}{4}\right)\frac{1}{\phi}\, e(T)_{O_2} + 3.76\left(n + \frac{m}{4}\right)\frac{1}{\phi}\, e(T)_{N_2}\right) \tag{9.6}$$

The internal energy $E_R(T)$ can be calculated from the polynomial coefficients given in Table 2.1. A numerical procedure which enables the internal energy for a mixture to be evaluated is given in Appendix II.A. The internal energy change E_2-E_1 (9.5) is then

$$E_2-E_1 = \left(E_R\right)_2 - \left(E_R\right)_1 = \left(E_{OR}\right)_2 - \left(E_{OR}\right)_1 + E_R\left(T_2\right) - E_R\left(T_1\right)$$

or

$$E_2-E_1 = E_R\left(T_2\right) - E_R\left(T_1\right), \tag{9.7}$$

since for a mixture of constant composition $\left(E_{OR}\right)_2 = \left(E_{OR}\right)_1$.

For a change in state from V_1 to V_2 the first law (9.3) is thus

$$E_2-E_1 + dW = 0. \tag{9.8}$$

Substituting for E_2-E_1 and dW we obtain

$$E_R\left(T_2\right) - E_R\left(T_1\right) + \left(\frac{p_1+p_2}{2}\right)\left(V_2-V_1\right) = 0. \tag{9.9}$$

The internal energy term $E_R(T_2)$ depends on T_2, therefore in expression (9.9) there are two unknowns p_2 and T_2; p_1, T_1 and $E_R(T_1)$ are known at the beginning of the step. These are related by the state equation

$$p_2 = p_1\left(\frac{V_1}{V_2}\right)\left(\frac{T_2}{T_1}\right). \tag{9.10}$$

To solve for p_2 and T_2 we use the Newton-Raphson method outlined in Chapter 8.

Let

$$f(E) = E_R(T_2)-E_R(T_1) + \left(\frac{p_1+p_2}{2}\right)(V_2-V_1) \tag{9.11}$$

and

$$f'(E) = \frac{df(E)}{dT} = \frac{d\left[E_R(T_2)\right]}{dT} = M_R C_v(T_2) \tag{9.12}$$

since from (2.13)

$$\frac{d\left[E_R(T_2)\right]}{dT} = M_R C_v(T_2) \tag{9.13}$$

and $$\frac{dE_R(T_1)}{dT} = 0.$$

In the expression (9.12) we have assumed that the rate of change of work W with temperature T over the volume change is negligible and

$$\frac{dW}{dT} = 0.$$

Solution of (9.11) is by Newton-Raphson. If $(T_2)_{n-1}$ is the estimated value of T_2, then

$$(T_2)_n = (T_2)_{n-1} - \frac{f(E)_{n-1}}{f'(E)_{n-1}}$$

or $$(T_2)_n = (T_2)_{n-1} - \frac{f(E)_{n-1}}{M_R C_v(T_2)} \tag{9.14}$$

The first estimate of T_2 can be obtained from the expression

$$T_2 = T_1\left(\frac{V_1}{V_2}\right)^{\kappa-1} = T_1\left(\frac{V_1}{V_2}\right)^{\frac{R_{mol}}{C_v(T_1)}}, \tag{9.15}$$

where $$\kappa = \frac{C_p(T_1)}{C_v(T_1)},$$

and the first estimated value of p_2 is obtained from (9.10). An algorithm for the step 1 to 2 is given in Fig. 9.2.

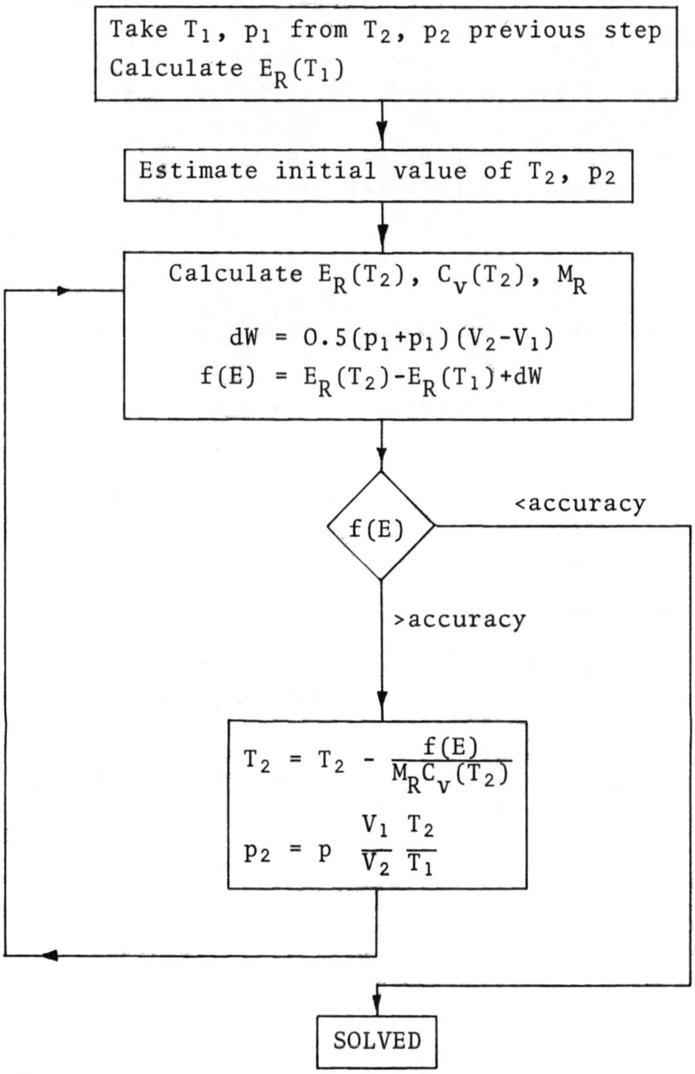

FIG. 9.2. Algorithm for compression stroke A to B.

9.1.2 Adiabatic Combustion at Constant Volume

This period corresponds to BC in the pressure and temperature diagrams (Fig. 9.1). In this simple analysis we shall assume the products of combustion to be carbon dioxide (CO_2), carbon monoxide (CO), water vapour (H_2O), hydrogen (H_2), oxygen (O_2) and nitrogen (N_2). Later we shall examine a more comprehensive list of products.

For the combustion of one mol of C_nH_m in air we shall have the following products:

$$a_1CO_2 + a_2CO + a_3H_2O + a_4H_2 + a_5O_2 + a_6N_2, \tag{9.16}$$

where a_1, a_2, a_3, a_4, a_5, and a_6 are the number of mols of CO_2, CO, H_2O, H_2, O_2 and N_2, respectively per mol of C_nH_m.

The general chemical reaction for the combustion process from B to C will then be of the form

$$w\left(C_nH_m + \left(n + \frac{m}{4}\right)\frac{1}{\phi}\,O_2 + 3.76\left(n + \frac{m}{4}\right)\frac{1}{\phi}\,N_2\right)$$

$$\longrightarrow w\left(a_1CO_2 + a_2CO + a_3H_2O + a_4H_2 + a_5O_2 + a_6N_2\right). \tag{9.17}$$

To determine the number of mols a_1 to a_6 we must consider the dissociation reactions (see Section 2.7). In the present case there are two such reactions:

$$CO + H_2O \rightleftharpoons H_2 + CO_2 \quad \text{and} \quad CO + \tfrac{1}{2}O_2 \rightleftharpoons CO_2.$$

The first reaction is the water-gas reaction: this states that in the equilibrium state the rate of formation of hydrogen (H_2) and carbon dioxide (CO_2) is equal to the rate of formation of carbon monoxide (CO) and water vapour (H_2O). The concentrations of CO, H_2O, H_2 and CO_2 in the equilibrium mixture can be determined from their partial pressures through the equilibrium equation (2.142); thus

$$\frac{P_{CO_2}\,P_{H_2}}{P_{CO}\,P_{H_2O}} = K_{P1}. \tag{9.18}$$

The partial pressures are the normalized partial pressures (2.140) and K_{P1} is the equilibrium constant.

The second reaction is the carbon monoxide (CO) reaction: this states that the rate of formation of carbon dioxide (CO_2) is equal to the rate of formation of CO and oxygen (O_2) at equilibrium. The concentrations of CO_2, CO, O_2 can be obtained from the equilibrium equation (2.139),

$$\frac{P_{CO_2}}{P_{CO}\sqrt{P_{O_2}}} = K_{P2}. \tag{9.19}$$

The equilibrium constants K_{P1}, K_{P2} can be determined from the Gibbs functions for the nominal reactants (subscript R) and products (subscripts P) through expression (2.144):

$$\ell n\ K_P = \Sigma\left(\frac{\nu g(T)}{R_{mol}T}\right)_R - \Sigma\left(\frac{\nu g(T)}{R_{mol}T}\right)_P - \frac{\Delta_{HO}}{R_{mol}T} \quad (9.20)$$

The specific Gibbs functions are determined from (2.145).

In Appendix II.A the numerical evaluation of the equilibrium constants K_{P1}, K_{P2} is given.

The heats of reaction (ΔH_0) are for the two reactions

$$CO + H_2O \longrightarrow H_2 + CO_2,$$

$$\Delta H_0 = -0.4047 \times 10^8 \text{ J/kg-mol}; \quad (9.21a)$$

$$CO + \tfrac{1}{2}O_2 \longrightarrow CO_2,$$

$$\Delta H_0 = -2.7969 \times 10^8 \text{ J/kg-mol}. \quad (9.21b)$$

The total number of mols of the products per mol of C_nH_m is

$$M_P = a_1 + a_2 + a_3 + a_4 + a_5 + a_6.$$

The normalized partial pressure P_i for substance i is

$$p_i = \frac{wa_i}{wM_P} P_C = \frac{a_i}{M_P} P_C,$$

where P_C is the normalized total pressure for the mixture at C (Fig. 9.1).

We can now substitute for the partial pressures P_i in the equilibrium equations (9.18) and (9.19) to give

$$\frac{a_1 a_4}{a_2 a_3} = K_{P1} \quad (9.22)$$

and

$$\left(\frac{a_1}{a_2}\right)^2 \frac{1}{a_5} = \frac{P_C}{M_P} K_{P2}^2 \quad (9.23)$$

To obtain the number of mols a_1 to a_6 we must prepare a chemical mass balance for carbon, hydrogen and oxygen. These are:

Carbon:

$$n = a_1 + a_2. \quad (9.24)$$

Hydrogen:

$$\frac{m}{2} = a_3 + a_4. \quad (9.25)$$

Oxygen:

$$2\left(n + \frac{m}{4}\right) \frac{1}{\phi} = 2a_1 + a_2 + a_3 + 2a_5. \qquad (9.26)$$

The state equation at C (Fig. 9.1) is

$$p_C V_C = w M_P R_{mol} T_C,$$

whence

$$\frac{p_C}{M_P} = \frac{w R_{mol} T_C}{V_C},$$

and in terms of the normalized pressure $P = p/p_O$ (2.140)

$$\frac{P_C}{M_P} = \frac{w R_{mol} T_C}{p_O V_C}. \qquad (9.27)$$

We now can combine (9.22) to (9.27) to give

$$f(a) = (a-B) - \frac{m}{2}\left(\frac{n-a}{n+Ca}\right) + \frac{2}{D}\left(\frac{a}{n-a}\right)^2 = 0, \qquad (9.28)$$

where

$$a = a_1, \qquad (9.29)$$

$$A = \frac{P_C}{M_P}, \qquad (9.30)$$

$$B = \frac{2}{\phi}\left(n + \frac{m}{4}\right) - \left(n + \frac{m}{2}\right), \qquad (9.31)$$

$$C = \frac{1}{K_{P1}} - 1, \qquad (9.32)$$

$$D = A\, K_{P2}^2. \qquad (9.33)$$

Expression (9.28) gives the number of mols of carbon dioxide ($a=a_1$) at the temperature T_C. The constant B is fixed from the fuel and equivalence. C, D and A are dependent on T_C, thus (9.28) can be solved for a given temperature T_C. Once a is known it follows from (9.29) that the number of mols a_1 of carbon dioxide are determined. We can then find the number of mols of carbon monoxide (a_2) from (9.24),

$$a_2 = m - a, \qquad (9.34)$$

and the number of mols of oxygen (a_5) from (9.23),

$$a_5 = \left(\frac{a_1}{a_2}\right)^2 \frac{M_P}{P_C} \frac{1}{K_{P2}^2} = \left(\frac{a_1}{a_2}\right)^2 \frac{1}{D} . \tag{9.35}$$

Combining (9.24), (9.25), (9.26) and (9.31) the number of mols of hydrogen (a_4) can be obtained from

$$a_4 = a_1 + 2a_5 - B \tag{9.36}$$

and the mols of water vapour (a_3) from (9.25),

$$a_3 = \frac{m}{2} - a_4 . \tag{9.37}$$

Finally, the nitrogen composition (a_6) is obtained directly from (9.17),

$$a_6 = \frac{3.76}{\phi} \left(n + \frac{m}{4}\right) . \tag{9.38a}$$

Thus if we know the temperature T_C we can determine the composition of the products of combustion. The temperature T_C is obtained from the first law,

$$dQ - dW = dE.$$

For adiabatic combustion $dQ = 0$ and for constant volume combustion $dW = 0$. It follows therefore that for the process from B to C,

$$dE = E_C - E_B = 0. \tag{9.38b}$$

As before, the internal energy E is defined as

$$E = E_0 + E(T).$$

At the state B the internal energy is

$$E_B = E_R = E_{OR} + E_R (T_B). \tag{9.39}$$

At the state C the internal energy is

$$E_C = E_P = E_{OP} + E_P (T_C). \tag{9.40}$$

We can express the internal energies in terms of the composition of the mixtures at B and C and the specific internal energies, we then obtain:

$$E_{OR} = w\left[\left(e_O\right)_{C_nH_m} + \left(n + \frac{m}{4}\right)\frac{1}{\phi}\left(e_O\right)_{O_2} + 3.76\left(n + \frac{m}{4}\right)\frac{1}{\phi}\left(e_O\right)_{N_2}\right], \tag{9.41}$$

$$E_R\left(T_B\right) = w\left[e\left(T_B\right)_{C_nH_m} + \left(n + \frac{m}{4}\right)\frac{1}{\phi}\,e\left(T_B\right)_{O_2} + 3.76\left(n + \frac{m}{4}\right)\frac{1}{\phi}\,e\left(T_B\right)_{N_2}\right], \tag{9.42}$$

$$E_{OP} = w\left[a_1\left(e_O\right)_{CO_2} + a_2\left(e_O\right)_{CO} + a_3\left(e_O\right)_{H_2O} + a_4\left(e_O\right)_{H_2} + a_5\left(e_O\right)_{O_2} + a_6\left(e_O\right)_{N_2}\right]. \tag{9.43}$$

$$E_P\left(T_C\right) = w\left[a_1\,e\left(T_C\right)_{CO_2} + a_2 e\left(T_C\right)_{CO} + a_3 e\left(T_C\right)_{H_2O} + a_4 e\left(T_C\right)_{H_2} + a_5 e\left(T_C\right)_{O_2} + a_6 e\left(T_C\right)_{N_2}\right]. \tag{9.44}$$

If we substitute for E_B and E_C into (9.38) we have

$$f(E) = \left(E_{OP} + E_P\left(T_C\right)\right) - \left(E_{OR} + E_R\left(T_B\right)\right) = 0. \tag{9.45}$$

Now the internal energies E_{OR}, $E_R\left(T_B\right)$ from (9.41) and (9.42) are fixed since they refer to the reactants at the temperature T_B. The internal energy E_{OP} will depend on the composition of the products and this depends on T_C; the internal energy $E_P\left(T_C\right)$ will depend on the composition of the products and the temperature. If we assume that $dE_{OP}/dT = 0$ we can write that

$$\frac{df(E)}{dT} = \frac{dE_P\left(T_C\right)}{dT} = E'_P\left(T_C\right) \tag{9.46}$$

and it follows from (2.13) that

$$E'_P\left(T_C\right) = M_C C_v\left(T_C\right) = wM_P C_v\left(T_C\right), \tag{9.47}$$

where

$$M_P = a_1 + a_2 + a_3 + a_4 + a_5 + a_6.$$

Equation (9.45) can then be solved numerically by Newton-Raphson's method:

$$\left(T_C\right)_n = \left(T_C\right)_{n-1} - \frac{f(E)_{n-1}}{M_C C_v \left(T_C\right)_{n-1}} . \tag{9.48}$$

The first estimate for T_C can be obtained by the approximate expressions due to Annand (Benson et al.[(1)]):

For $\phi < 1.0$ $T_C = T_B + 2500\,\phi.$

For $\phi > 1.0$ $T_C = T_B + 2500\,\phi - 700\,(\phi-1).$

To calculate the temperature at the end of combustion T_C we must first evaluate the composition and then test the first law energy balance (9.45). A suitable algorithm is given in Fig. 9.3. The equilibrium constants K_p can be calculated from (9.20) and (9.21) by the methods outlined in Appendix II.A.

The maximum temperature in the cycle corresponds to T_C. This is dependent on the air/fuel ratio, the initial temperature and pressure at the commencement of combustion and the fuel. We shall examine this later.

9.1.3 Adiabatic Expansion

This corresponds to the period from C to D in the pressure and temperature diagram.

Although during combustion the maximum temperature is reached, the chemical reactions continue to take place during the expansion stroke; this is because the chemical species are in thermodynamic equilibrium. The composition of the gas mixture will therefore vary with pressure and temperature. To satisfy thermodynamic equilibrium the chemical reactions for water gas and carbon dioxide, namely,

$$CO + H_2O \rightleftharpoons H_2 + CO_2 \qquad CO + \tfrac{1}{2}O_2 \rightleftharpoons CO_2,$$

are continuously taking place. The analysis for the determination of the composition given for the combustion process (Section 9.1.2) is therefore the same for the expansion process. If we consider a step from V_1 to V_2, then the pressure and temperature will change from p_1, T_1 to p_2, T_2.

These will be related by

$$p_1 V_1 = M_1 R_{mol} T_1 \quad \text{and} \quad p_2 V_2 = M_2 R_{mol} T_2,$$

and the number of mols at 1 and 2 will be

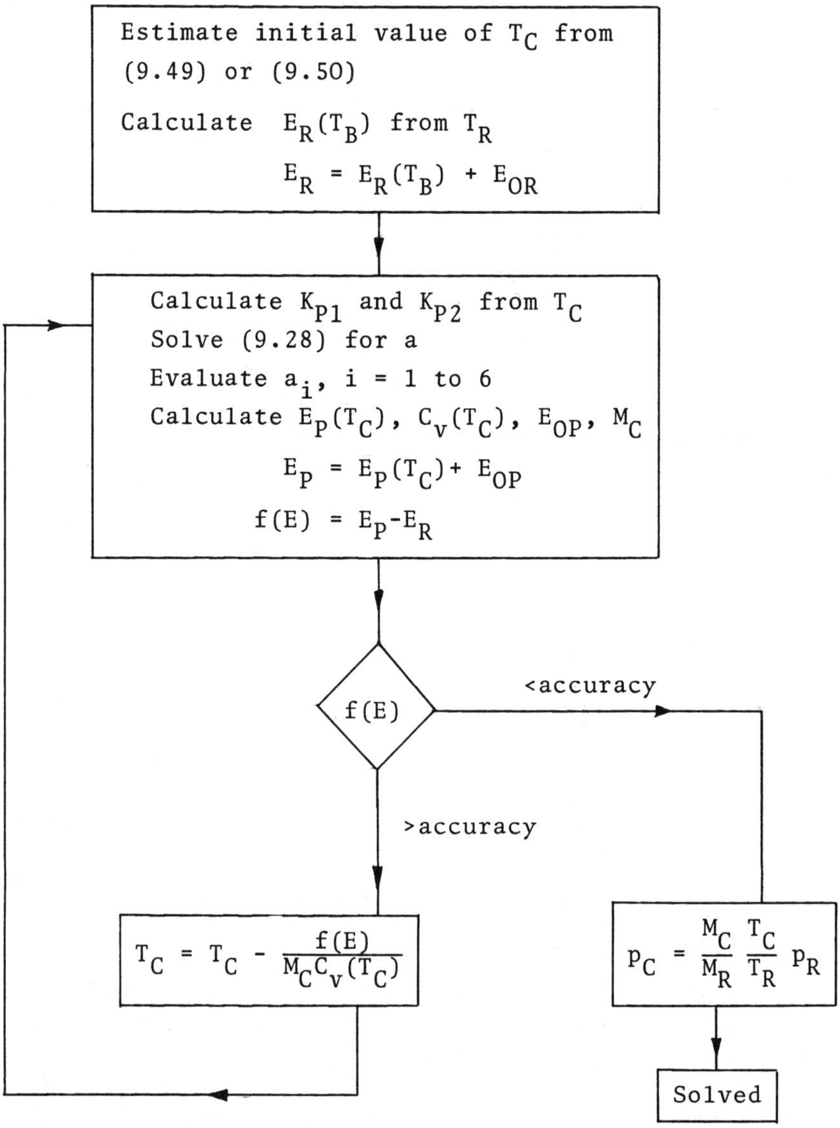

FIG. 9.3. Algorithm for constant volume combustion B to C.

$$M_1 = w\left(a_1 + a_2 + a_3 + a_4 + a_5 + a_6\right)_{T_1},$$

$$M_2 = w\left(a_1 + a_2 + a_3 + a_4 + a_5 + a_6\right)_{T_2}.$$

Instead of P_C/M_P in (9.27) and (9.30), we use

$$A = \frac{P_2}{M_2} = \frac{wR_{mol}T_2}{p_0V_2} \tag{9.49}$$

Otherwise expressions (9.31), (9.32), (9.33) are only dependent on T_2. Hence if T_2 is known we can calculate a_1 to a_6.

To obtain T_2 we apply the first law,

$$dQ - dW = dE. \tag{9.50}$$

For an adiabatic process $dQ = 0$ and for a small step V_1 to V_2 the work is approximately

$$dW = p\ dV = \frac{p_1+p_2}{2}\left(V_2-V_1\right). \tag{9.51}$$

The change in internal energy dE is

$$dE = E_2 - E_1, \tag{9.52}$$

where, as before,

$$E = E_P = E_{OP} + E_P(T). \tag{9.53}$$

For temperature T_1 the internal energies are:

$$E_{OP1} = w\Big(a_1\left(e_O\right)_{CO_2} + a_2\left(e_O\right)_{CO} + a_3\left(e_O\right)_{H_2O}$$

$$+ a_4\left(e_O\right)_{H_2} + a_5\left(e_O\right)_{O_2} + a_6\left(e_O\right)_{N_2}\Big), \tag{9.54}$$

$$E_P\left(T_1\right) = w\Big(a_1e\left(T_1\right)_{CO_2} + a_2e\left(T_1\right)_{CO} + a_3e\left(T_1\right)_{H_2O}$$

$$+ a_4e\left(T_1\right)_{H_2} + a_5e\left(T_1\right)_{O_2} + a_6e\left(T_1\right)_{N_2}\Big), \tag{9.55}$$

and at T_2:

$$E_{OP} = w\Big(a_1\left(e_O\right)_{CO_2} + a_2\left(e_O\right)_{CO} + a_3\left(e_O\right)_{H_2O}$$

$$+ a_4\left(e_O\right)_{H_2} + a_5\left(e_O\right)_{O_2} + a_6\left(e_O\right)_{N_2}\Big), \tag{9.56}$$

$$E_P\left(T_2\right) = w\left[a_1 e\left(T_2\right)_{CO_2} + a_2 e\left(T_2\right)_{CO} + a_3 e\left(T_2\right)_{H_2O}\right.$$

$$\left. + a_4 e\left(T_2\right)_{H_2} + a_5 e\left(T_2\right)_{O_2} + a_6 e\left(T_2\right)_{N_2}\right], \qquad (9.57)$$

the number of mols a_1 to a_6 being different at each temperature. At the initial temperature T_1 the internal energies E_{OP1} and $E_P(T_1)$ are fixed; however, E_{OP2} will not necessarily be equal to E_{OP1} due to the change in composition of the mixture. The first law (9.50) then becomes

$$f(E) = \left(E_{OP2} + E_P\left(T_2\right)\right) - \left(E_{OP1} + E_P\ T_1\right)\Big)$$

$$+ \left(\frac{p_1 + p_2}{2}\right)\left(V_2 - V_1\right) = 0. \qquad (9.58)$$

If we let $\frac{dW}{dT} = 0$ and $\frac{dE_{OP}}{dT} = 0,$

then, as before,

$$\frac{df(E)}{dT} = \frac{dE_P\left(T_2\right)}{dT} = E'_P\left(T_2\right) = M_2 C_V\left(T_2\right). \qquad (9.59)$$

We can solve expression (9.58) numerically by Newton-Raphson:

$$\left(T_2\right)_n = \left(T_2\right)_{n-1} - \frac{f(E)}{M_2 C_V\left(T_2\right)_{n-1}} \qquad (9.60)$$

The first estimate for T_2 is the same as (9.15), namely,

$$T_2 = T_1\left(\frac{V_1}{V_2}\right)^{\frac{R_{mol}}{C_V\left(T_1\right)}} \qquad (9.61)$$

and the pressure ratio is

$$\frac{p_2}{p_1} = \frac{M_2}{M_1}\frac{V_1}{V_2}\frac{T_2}{T_1}. \qquad (9.62)$$

A suitable algorithm for the calculation of T_2 is given in Fig. 9.4. Notice the calculation involves first estimating T_2 and calculating the composition of the mixture followed by the check using the first law of thermodynamics.

To close the cycle we join p_1 to p_4 at constant volume.

A computer program based on the above analysis is given in Appendix II.C.

9.1.4 Cycle Studies

The indicated thermal efficiency is the ratio of the work done to the energy content of the fuel supplied. The work done per unit time is the area of the pressure volume diagram $\left(\oint_A^A p\,dV\right)$ times the number of cycles N. If the heat of reaction is Q_{VS} at the standard reference temperature and the fuel rate is $\dot{m}_f$ units of mass per unit time, then we can define the indicated thermal efficiency as

$$\eta_{TH} = \frac{\oint_A^A p\,dV\,N}{-\dot{m}_f Q_{VS}} \quad \text{or} \quad \frac{\oint_A^A p\,dV}{-wQ_{VS}}.$$

The negative sign is introduced since the heat of reaction Q_{VS} will be negative. The mean effective pressure p_m is the area of the indicated diagram divided by the swept volume V_S. Hence

$$p_m = \frac{\oint_A^A p\,dV}{V_S}.$$

In Fig. 9.5 the indicated thermal efficiency, indicated mean effective pressure and peak temperature are shown over a range of air/fuel ratios for an ideal Otto cycle with octane-air mixture and compression ratio of 7.0.

The maximum temperature varies with the air/fuel ratio. If dissociation were not present, the peak temperature would occur at the stoichiometric mixture since the reaction would go to completion; however, due to dissociation and the influence of the specific heats of carbon monoxide and carbon dioxide, the maximum temperature is on the rich side. This can be explained by the diagram shown in Fig. 9.6, which is not to scale. The effect of dissociation is illustrated in Fig. 9.6(a), where the temperature is reduced from T_{C1} to T_{C2} due to the dissociation energy "loss" $x\Delta E$. On the other hand, the presence of carbon monoxide, which has a lower specific internal energy, and hence lower specific heat, than carbon dioxide produces an increase in temperature from T_{C2} to T_{C3}, as shown in

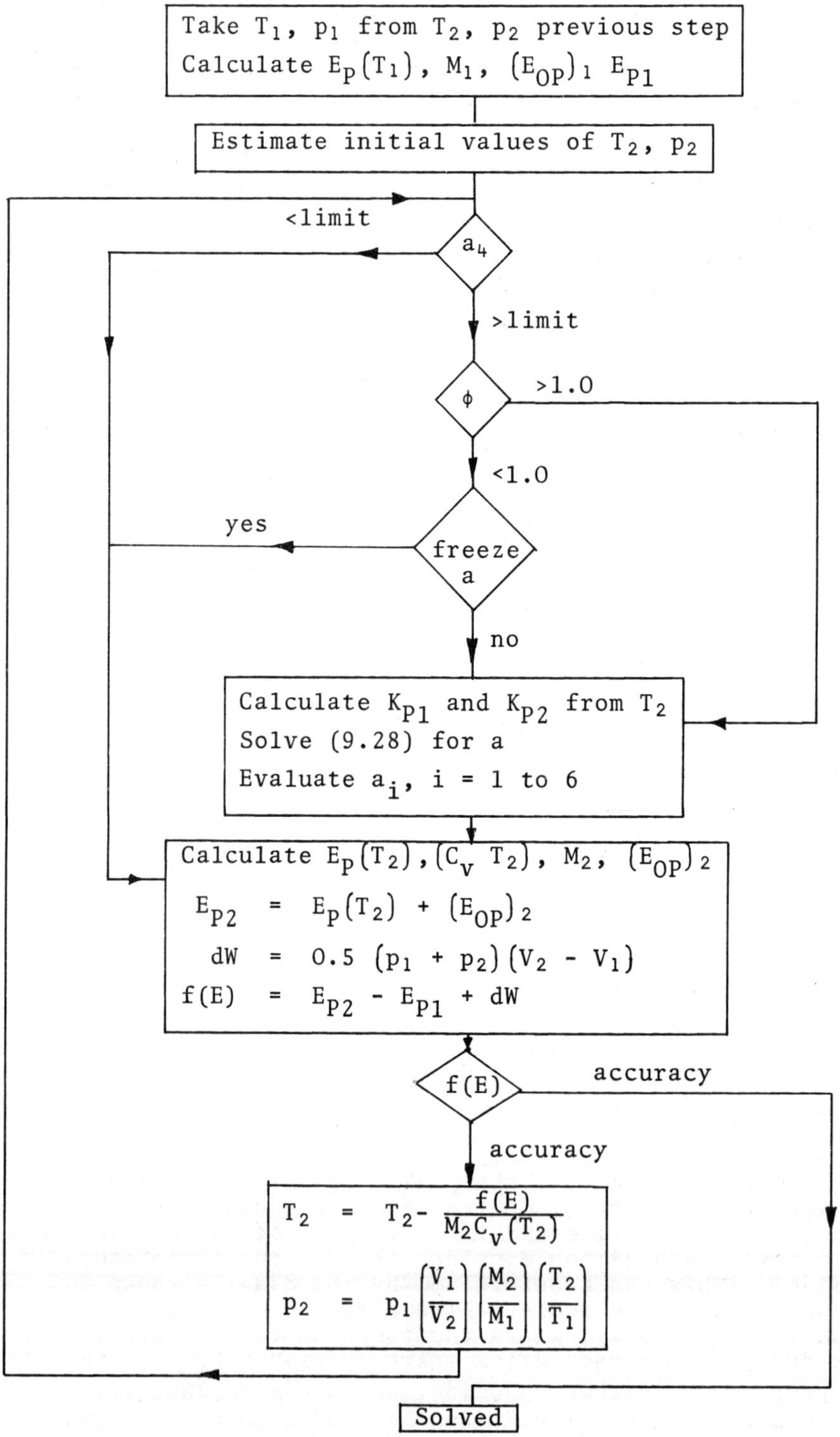

FIG. 9.4. Algorithm for expansion stroke C to D.

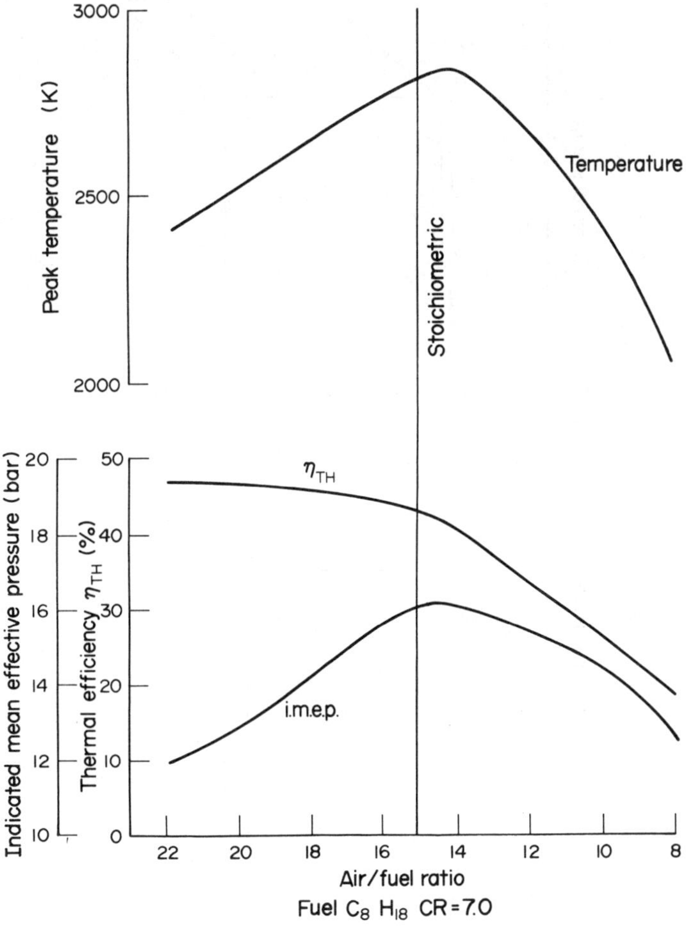

FIG. 9.5. Otto cycle performance—effect of fuel/air ratio.

Fig. 9.6(b). The combination of these effects give the results shown in Fig. 9.5 for the variation of peak temperature with air/fuel ratio.

Excepting for nitrogen the concentration of all the gases varies throughout the expansion stroke, as will be seen in Fig. 9.7. The most significant changes are in the concentrations of carbon monoxide (CO) and carbon dioxide (CO_2). In Fig. 9.8 the CO and CO_2 concentrations at the peak temperature and at the end of the expansion stroke are shown for a range of air/fuel ratios. Here it will be seen that CO is present at the peak temperature even with extremely weak mixtures (22:1), but disappears at the end of the expansion stroke when the mixture is weak. The difference between the CO concentration at the beginning and the end of the expansion stroke reaches a minimum at the maximum temperature; at the same

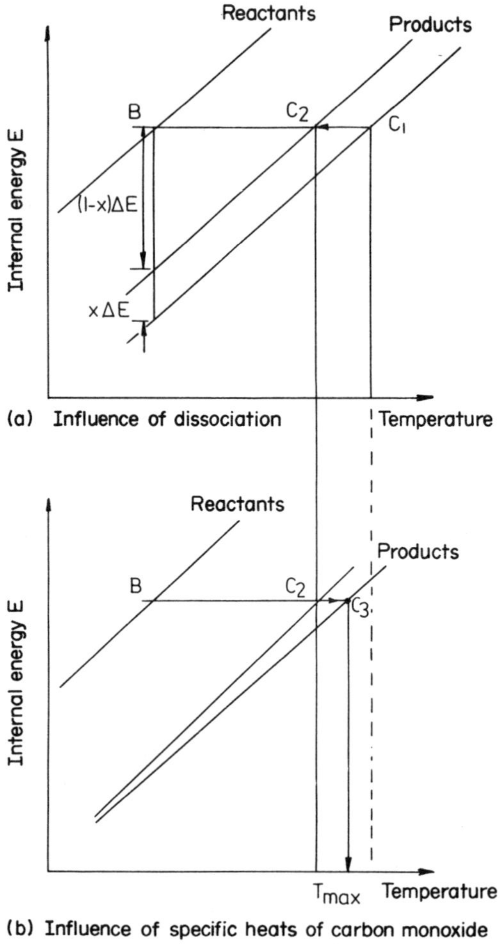

FIG. 9.6. Diagrammatic representation of the effect of dissociation and variable specific heats on maximum temperature.

air/fuel ratio there is the minimum difference between the CO concentration at the peak temperature and at the end of the expansion stroke. At this air/fuel ratio the two effects referred to in Fig. 9.6 are at their optimum relationship, ensuring the maximum temperature rise during combustion. The mean effective pressure is at a maximum at the same air/fuel ratio as the peak temperature; on the other hand, the thermal efficiency increases with increasing weakness rapidly up to about the air/fuel ratio corresponding to the maximum temperature, then it increases at a lower rate. However, even at the weakest mixture the thermal efficiency is less than the air standard efficiency, as will be seen from Fig. 9.9.

On the rich side the decrease in thermal efficiency with increase in fuel flow is due to the influence of the excess fuel on the CO concentration and hence the unreleased energy (Fig. 9.8). On the

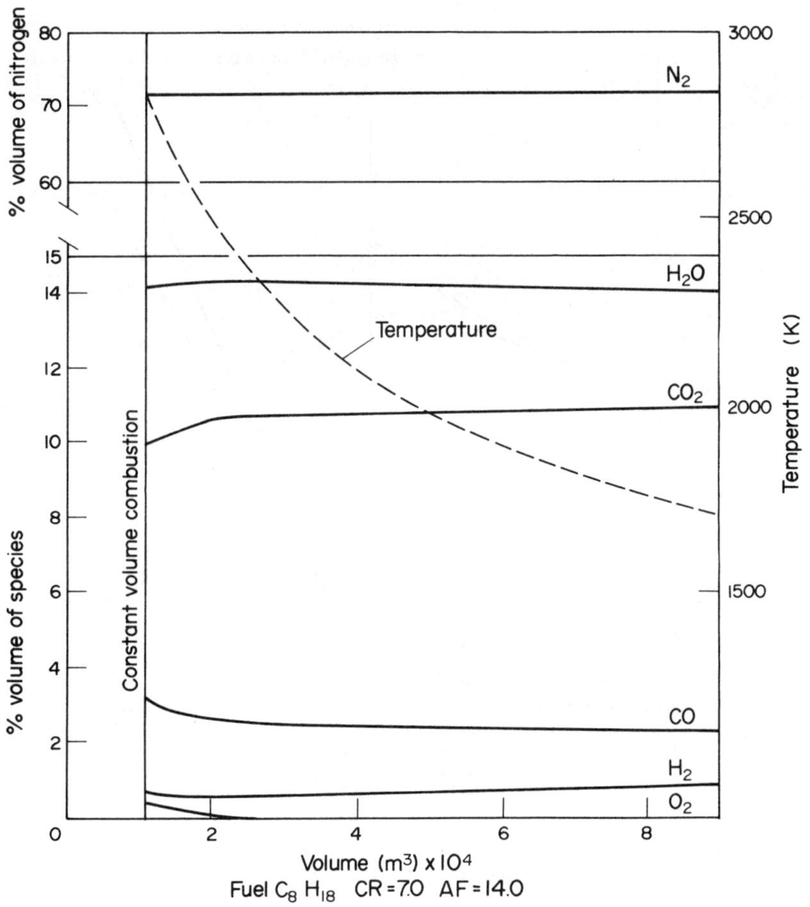

FIG. 9.7. Variation of composition in expansion stroke.

weak side the variation in thermal efficiency is due to the influence of the specific heats of the products of combustion. The higher the cycle temperatures, the higher the specific heats and hence the lower the temperature rise during combustion and the lower the efficiency. A second influence is the presence of the triatomic gas CO_2 in the products which also raises the specific heats and hence limits the temperature rise. If we examine variations of the thermal efficiency and the inverse of the thermal efficiency with the mean effective pressure, we obtain the double branch curves shown in Fig. 9.10. The inverse of the thermal efficiency is proportioned to the specific fuel consumption (kg/kWh). The lower curve has the form of a typical fuel consumption loop for a gasolene engine. Thus as we increase the fuel from the point A there is a slow increase in specific fuel consumption until we reach maximum power, then further increase in fuel reduces the power with a rapid increase in specific fuel consumption.

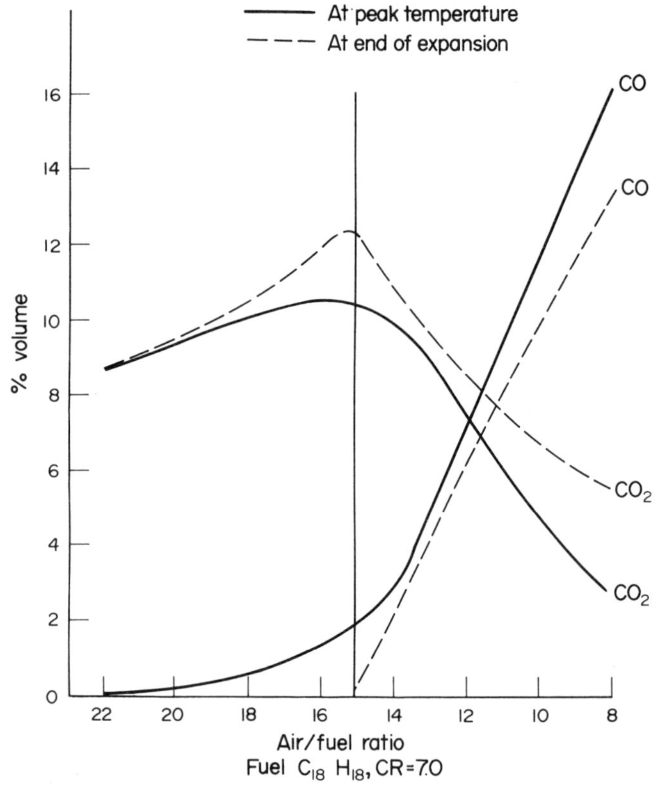

FIG. 9.8. Composition at peak temperature and end of expansion.

The combination of maximum temperature rise and the thermal efficiency gives the maximum power at a rich mixture. Thus dissociation moves the thermodynamics point for production of maximum power to a region of lower thermal efficiency based on fuel consumption. If we plot the i.m.e.p. against peak oxygen concentration at maximum temperature (Fig. 9.11), we clearly see there is still over 1% oxygen present at the stoichiometric ratio. This is reduced to about 0.5% at maximum power. It could be argued that the limiting criterion influencing the thermodynamics conversion of chemical energy to work energy is the oxygen and not the hydrocarbon. That is, provided there is oxygen present useful work can be produced. Pye(2) refers to the thermal efficiency based on air consumption. For a full discussion on this approach to the thermodynamic efficiency the reader is referred to Pye(2).

The fuel composition can also influence the peak temperature and power, as will be seen in Fig. 9.12. Propane (C_3H_8) has a higher heat of reaction than octane (C_8H_{18}); at the same time the composition of the products are different, there being a slightly

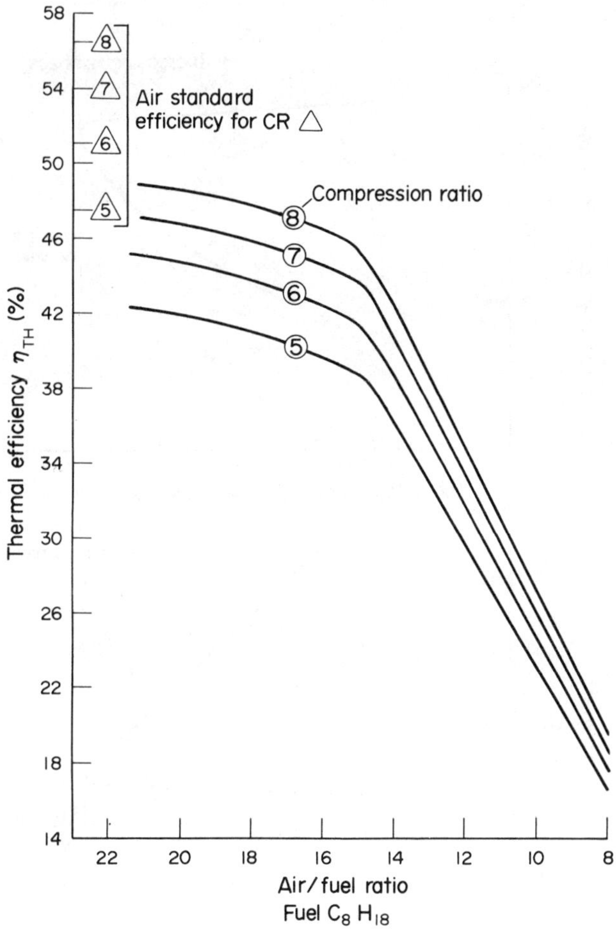

FIG. 9.9. Influence of compression ratio on thermal efficiency.

higher concentration of CO_2 and CO with octane which will influence the specific heats of the products. These two effects give a higher temperature rise and power for propane than for octane despite the slightly higher thermal efficiency for octane over most but not all of the air/fuel ratio range.

The cycle performance we have analysed refers to an ideal Otto cycle. In practice the finite time for combustion and heat losses will influence the overall cycle predictions. We shall now examine a more realistic model for a spark ignition engine cycle.

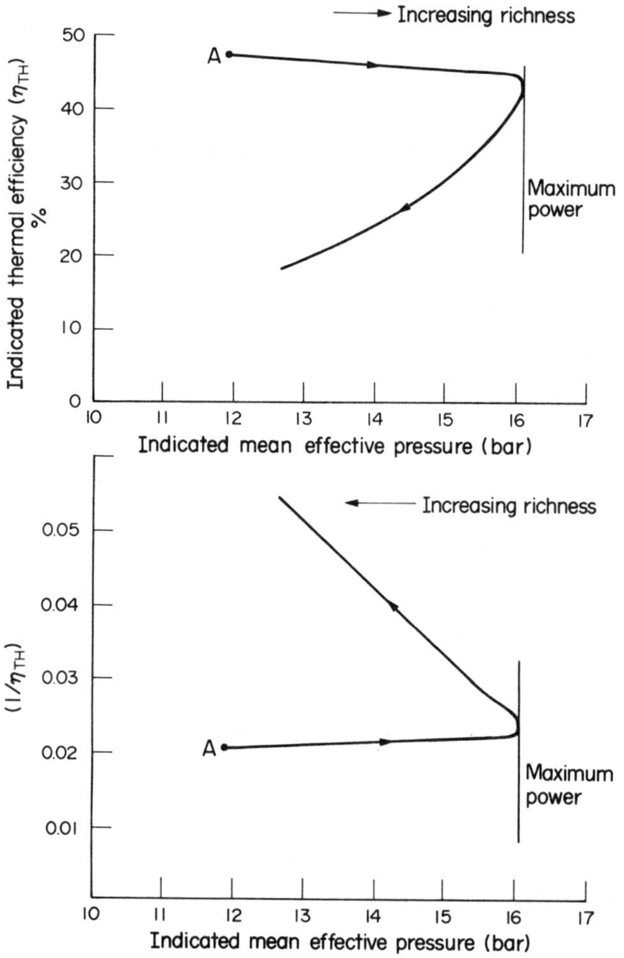

FIG. 9.10. Variations of thermal efficiency with imep.

9.2 CYCLE CALCULATIONS WITH ALLOWANCE FOR COMBUSTION TIME, HEAT LOSS AND RATE KINETICS

Detailed calculations for real engines are complex. Space precludes a full development of the methods used; the reader is referred to the paper by Benson et al.[1] for a full treatment. In this section an outline of the method is given.

The basic principles are the same for a real cycle calculation as for an ideal cycle except that allowance must be made for heat transfer and flame travel times. The major assumptions are that

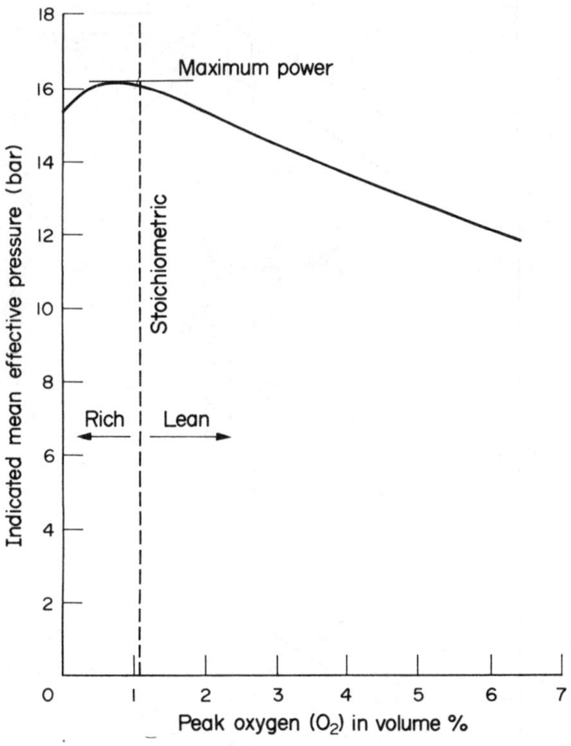

FIG. 9.11 Power versus peak oxygen concentration.

the mixture is homogeneous, the flame front is thin and that during combustion the combustion chamber is subdivided into two zones—a burnt zone behind the flame front and an unburnt zone ahead of the flame front. At all times the pressure is uniform throughout the cylinder during combustion; however, there may be two temperatures—one behind the flame front (T_b) and the other ahead (T_a). Whilst heat transfer is considered from the cylinder to the walls, heat transfer between gas zones is not.

Numerically, the calculation techniques are different from the ideal cycles and the compression ignition engine cycles discussed earlier. The first law and the state equations are expressed in differential form, the variables being the angle α and the derivations $dP/d\alpha$, where P is the property. Runge-Kutta[3] methods are used with single-zone regions and simple Euler methods for two-zone regions. This implies that the method of integration changes in the calculation during the combustion period. Except for the initiation of combustion the internal energy is given in the form $de = C_V(T)dT$, where $C_V(T)$ is the specific heat at constant volume at temperature T derived from the gas composition and the temperature using the thermodynamics of Table 2.1. For a small time step $\Delta\alpha$ the specific heats are considered to be constant and correspond to the temperature T during the time step. We then have

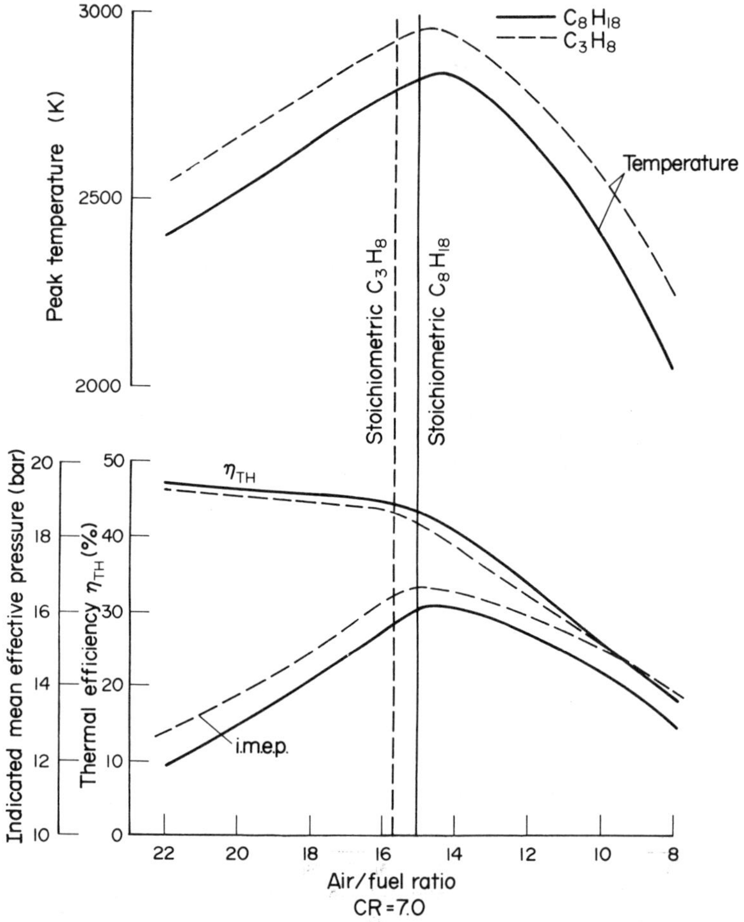

FIG. 9.12. Influence of fuel composition on performance.

$$\kappa(T) = \frac{C_p(T)}{C_v(T)}$$

and the normal isentropic relations are used over the time step. If the time step is small the above approximation gives accurate results.

The heat transfer expression of the conventional form (Chapter 6) are used. The calculations of pressure and temperature and composition are carried out simultaneously. The composition in the burnt zones is considered to contain twelve species H_2O, H_2, OH, H, N_2, NO, N, CO_2, CO, O_2, O, A. Here you will observe the inclusion of the atomic forms of hydrogen, nitrogen and oxygen as well as oxides of nitrogen. The following seven equilibrium reactions are considered:

$$\tfrac{1}{2}H_2 \rightleftharpoons H, \qquad H_2O \rightleftharpoons OH + \tfrac{1}{2}H_2,$$

$$\tfrac{1}{2}O_2 \rightleftharpoons O, \qquad CO_2 + H_2 \rightleftharpoons H + CO,$$

$$N_2 \rightleftharpoons N, \qquad H_2O + \tfrac{1}{2}N_2 \rightleftharpoons H_2 + NO.$$

$$2H_2O \rightleftharpoons 2H_2 + O_2,$$

The temperature is calculated from the derivative of temperature in the previous time step $dT/d\alpha$. With the new temperature the equilibrium proportions of the constituents are calculated from the dissociation equations for the above seven reactions by the methods of Vickland et al.[5] To calculate the nitric oxide (NO) concentration rate, controlled equations are used. The method is based on the so-called quasi-stationary assumptions of Lavoie,[6] so that one differential equation can be set up for NO only. In this assumption the actual concentrations of (NO), (N) and (N_2O) are related to the equilibrium values $(NO)_e$, $(N)_e$, (N_2O) by expressions of the form discussed in Chapter 2 (equation (2.160)).

The sequence of reactions forming N, NO and NO_2 are given by Annand[11] as:

(1) $N + NO \rightleftharpoons N_2 + O.$

(2) $N + O_2 \rightleftharpoons NO + O.$

(3) $N + OH \rightleftharpoons NO + H.$

(4) $H + N_2O \rightleftharpoons N_2 + OH.$

(5) $O + N_2O \rightleftharpoons N_2 + O_2.$

(6) $O + N_2O \rightleftharpoons NO + NO.$

(7) $N_2O + M \rightleftharpoons N_2 + O + M.$

Lavoie[6] states that the reaction times for the formation of atomic nitrogen and nitrous oxide are several orders of magnitude faster than for NO so that the significant constituent is NO. Using the methods outlined in Chapter 2 (and the notation in (2.157)) it can be shown[1] that the rate of formation of NO is

$$\frac{1}{V}\frac{d((NO)V)}{dt} = 2(1-\alpha^2)\left(\frac{R_1}{1+\alpha\dfrac{R_1}{(R_2+R_3)}} + \frac{R_6}{1+\dfrac{R_6}{R_4+R_5+R_7}}\right), \tag{9.63}$$

where $$\alpha = \frac{(NO)}{(NO)_e},$$

$$R_i = k_{fi}(A)_e(B)_e = k_{bi}(C)_e(D)_e, \tag{9.64}$$

and k_{fi} is the forward reaction rate constant for the above reactions, i = 1 to 7, in m^3/kg-mol/s and k_{bi} the corresponding backward reaction rate constant. V is the volume of the burnt gas $(= V_b)$ and T is the temperature $(= T_b)$.

The calculation procedure starts with the trapped mass of fuel, air and residuals. The pressures and temperatures in the compression stroke are then calculated, using the first law, until the nominal spark time, when the combustion period is said to commence. The calculation then proceeds in three phases. Firstly, the initiation of combustion, then the subdivision of the combustion chamber into two zones separated by a spherical flame front and, finally, a single zone encompassing the whole of the combustion chamber.

To initiate combustion a unit mass of the cylinder contents is considered to burn at constant volume. The internal energy of the initial reactants are set equal to the internal energy of the products and the temperature of the products computed in the same manner as in the ideal cycle. From the unburnt mixture temperature T_u and the products temperature T_b the laminar flame speed u is calculated from Kuehl's(4) expression or some similar equation. Kuehl's expression is

$$u_\ell = \left[\frac{1.087 \times 10^6}{\left(\left(\frac{10^4}{T_b}\right) + \left(\frac{900}{T_u}\right)\right)^{4.938}} \right] p^{-0.0987} \text{ cm/s}, \tag{9.65}$$

where p is the pressure in inches of mercury and the temperatures are in ^oK.

The turbulent flame speed is then calculated from

$$u_t = ff\, u_\ell, \tag{9.66}$$

where ff is a flame factor. Kuehl's expression is an empirical relationship based on a propane-air mixture. Experiments show that the flame factor ff is a constant related to engine design and speed.(1)

The radius of the burnt zone r, centred on the spark point, is given by

$$r = \frac{u_t\, \Delta\alpha}{360\ N},$$

where N is the engine speed in rev/s and $\Delta\alpha$ the crankangle step in degrees.

The corresponding volume of the burnt mixture is

$$V_p = \frac{2}{3} \pi r^3. \tag{9.67}$$

For initiation of combustion V_p is set equal to 0.1% of the total cylinder volume V_c. If V_p is less than this value, combustion is said not to have been initiated. Hence the delay period is

$$(\Delta\alpha)_{delay} = \left(\frac{360N}{u_t}\right)\left(\frac{0.0015V_c}{\pi}\right)^{\frac{1}{3}} . \quad (9.68)$$

If the total time step Δ from the nominal ignition of the spark is less than $(\Delta\alpha)_{delay}$ the compression stroke is continued. Once $\Delta\alpha$ measured from the nominal spark time is greater than $(\Delta\alpha)_{delay}$ we enter the second phase of the combustion calculation when the cylinder is divided into two zones separated by a spherical flame front centred on the spark plug.

To initiate the two-zone calculation the temperatures in the burnt and unburnt zones must be adjusted to ensure uniform pressure distribution in the cylinder. We will use subscripts u and b to represent the initial values and subscripts, uf and bp the final values for the unburnt and burnt zone pressures and temperatures, respectively. The procedure is as follows: a unit mass of the cylinder contents is considered. The temperature T_b is calculated from the first law, allowing for heat loss and dissociation from the initial pressure and temperature of the unburnt mixture p_u and T_u. This will give a second pressure p_b for the burnt gases from the expression

$$p_b = \left(\frac{M_b}{M_u}\right)\left(\frac{T_b}{T_u}\right) p_u, \quad (9.69)$$

where M_b and M_u are the number of mols per unit mass for the products and initial reactant, respectively. From the volume of the initial nucleus of the burnt zone V_b the burnt mass m_b can be calculated from p_b and T_b. From the total mass in the cylinder m_c the unburnt mass $m_u = m_c - m_b$ can be determined, and hence the internal energies of the cylinder contents obtained, thus,

$$E = m_b e_b + m_u e_u . \quad (9.70)$$

We now have the pressures in the two zones p_b and p_u, which from (9.69) are clearly not equal. These are equalized by assuming isentropic changes in temperature to an equalized pressure $p_c = p_{bf} = p_{uf}$. The basic relations are:

Burnt zone:

$$\frac{T_{bf}}{T_b} = \left(\frac{p_c}{p_b}\right)^{\frac{\kappa_b - 1}{\kappa_b}} \quad (9.71)$$

Unburnt zone:

$$\frac{T_{uf}}{T_u} = \left(\frac{p_c}{p_u}\right)^{\frac{\kappa_u - 1}{\kappa_u}}, \qquad (9.72)$$

where κ_b and κ_u are the ratios of the specific heats in the burnt and unburnt zones assumed constant over the step $\Delta\alpha$.

A third relationship is required between the temperatures and this is the energy balance

$$m_u(C_v)_u(T_{uf}-T_u) = m_b(C_v)_b(T_{fb}-T_b). \qquad (9.73)$$

Equations (9.71) to (9.73) provide the complete solution. Full details of the calculation technique are given in reference 1.

With the pressure and temperatures in each zone defined, the appropriate derivatives with respect to angle can be evaluated from the time step $\Delta\alpha$. From the temperature in the burnt zone the nitric oxide concentration is calculated. Using the derivatives $dT_b/d\alpha$, $dT_u/d\alpha$ the new temperatures in the two zones can be calculated for the next time step. Then the flame speed followed by the new volume of the burnt zone. Then the new constituents in the burnt zone, from the dissociation data; the specific heats of the two zones from the thermodynamic data; the nitric oxide concentration in the burnt zone from (9.63); and, finally, using the first law in differential form, the derivatives of temperature in the burnt and unburnt zones and the derivative of cylinder pressure can be determined. The calculation is repeated for each time step until the flame encloses the whole of the cylinder contents and we have a single burnt zone. By this time the piston is moving downwards and we are in the expansion stroke. The calculation procedure is now simplified and is very similar to the compression stroke except that for each time step the proportions of the constituents must be determined from equilibrium data and the nitric oxide concentration evaluated. Depending on the pressure and temperature in the cylinder the nitric oxide concentration will become almost a constant volume. It is then said to freeze. At this point the nitric oxide rate equation may be by-passed.

The calculation of the burnt zone is complex due to the interaction of the spherical flame front and the cylinder head, liner and piston surfaces. Approximate expressions for the volume and surface area for a disc type combustion chamber are given by Annand.[7]

It will be clear that cycle calculations with allowance for combustion time, heat losses and rate kinetics are extremely complex. In this chapter we have outlined the basic thermodynamics and chemical kinetics. For details of the alternative models the reader is referred to papers by Lavoie[6] and Blumberg.[10] In Chapter 5 the results of calculations using the methods outlined in this chapter are shown in Figs. 5.13 and 5.18. In Fig. 9.13 a comparison is made of the cycle predictions based on the ideal Otto cycle methods outlined in the first part of this chapter and the real cycle calculation in the second part.

It will be seen that the trend for the maximum temperature T_{max} is the same for the Otto cycle and the real cycle. On the other hand, there is a clear difference in the thermal efficiency and power. These differences are due to the combustion process taking a finite time in the real cycle and to the heat losses.

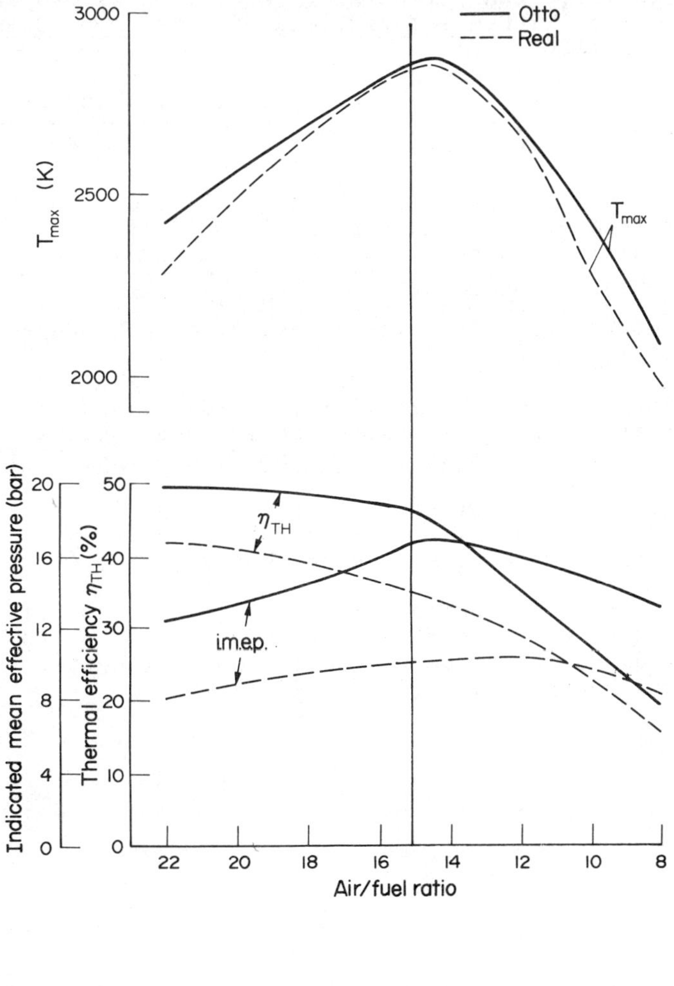

FIG. 9.13. Comparison of real and Otto cycle calculations.

REFERENCES

1. Benson, R.S., Annand, W.J.D. and Baruah, P.C., A simulation model including intake and exhaust systems for a single cylinder four-stroke cycle spark ignition engine, Int. J. Mech. Sci. 17 (2) 97 (1975).

2. Pye, D.R., The Internal Combustion Engine, Vol. 1, Oxford University Press, 1937.

3. James, M.L., Smith, G.M. and Wolford, J.C., Applied Numerical Methods for Digital Computation with FORTRAN, International Text Book Co., 1967.

4. Kuehl, D.K., Laminar-burning velocities of propane air mixtures, Eighth International Symposium on Combustion, p. 510 (1962).

5. Vickland, C.W., Strange, F.M., Bell, R.A. and Starkman, E.S., A consideration of the high temperature thermodynamics of internal combustion engines, SAE Trans. 170, 785 (1962).

6. Lavoie, G.A., Heywood, J.B. and Keck, J.C., Experimental and theoretical study of nitric oxide formation in internal combustion engines, Combust. Sci. and Technol. 1, 313 (1970).

7. Annand, W.J.D., Research notes, J. Mech. Engng. Sci. 12, 146 (1970).

8. Blumberg, P. and Kummer, J.T., Prediction of nitric oxide formation in spark ignited engines—an analysis of methods of control, Combust. Sci. and Technol. 4, 73 (1971).

9. Annand, W.J.D., Effects of simplifying kinetic assumptions in calculating nitric oxide formation in spark ignition engines, Proc. Instn. Mech. Engrs., 188, 431 (1974).

10. Benson, R.S. and Baruah, P.C., A generalised calculation for an ideal Otto cycle with hydrocarbon-air mixture allowing for dissociation and variable specific heats, Int. J1. Mech. Engng. Ed. 4 (1), 49-81 (1976).

Chapter 10
Supercharging

Notation

a	speed of sound
C_p	specific heat at constant pressure
C_v	specific heat at constant volume
C_{TS}	isentropic speed
D	diameter
E	energy
h	enthalpy
m	mass
n	rotational speed or polytropic index
N	rotational speed
p	pressure
p_m	mean effective pressure
p_R	reference pressure
P	power
Q	heat transfer
R	gas constant
T	temperature
V	volume
W	work
W_S	rotational speed or particle velocity
ρ	density
ϕ	equivalence
λ	scavenge ratio or Riemann variable (Section 10.8.2)
β	Riemann variable (Section 10.8.2)
κ	isentropic index, ratio of specific heats
ζ_L	proportion of fuel energy lost in cooling system
ζ_E	proportion of fuel energy in exhaust
Δh_{is}	isentropic enthalpy drop
η_A	exhaust pipe efficiency
η_C	adiabatic (isentropic) efficiency
η_M	mechanical efficiency
η_{SC}	scavenge efficiency
η_T	average turbine efficiency
η_{TC}	turbocharger efficiency
η_{TH}	thermal efficiency
η_{TS}	instantaneous total to static turbine efficiency

General subscripts

a	air
C	compressor
E	engine or exhaust
f	fuel
g	gas
ref	reference
T	turbine
01	stagnation inlet to compressor
02	stagnation outlet from compressor or inlet to engine
03	stagnation in exhaust pipe at entry to turbine
04	stagnation at exit from turbine

As we have seen, an internal combustion engine converts some of the energy released during combustion process to useful work; however, there is an upper thermodynamic limit to the efficiency of conversion and a practical limit to the degree of success in approaching the thermodynamic efficiency. The power produced by an engine is approximately proportional to the quantity of fuel burnt, and the latter is controlled by the air supplied to the engine. In naturally aspirated four-stroke engines the quantity of air is proportional to the volume swept by the piston (or some percentage in excess in two-stroke engines). In order to increase the air supply an external air source is required. When this is provided the engine is said to be supercharged. Supercharging has been applied to both spark ignition and compression ignition engines, but the latter application is in more general use and we shall direct our discussions to these engines. In this chapter we shall mainly confine our studies to the thermodynamics of supercharging processes, but we shall discuss some of the gas dynamics and flow considerations. We shall first discuss the relationship between the trapped conditions and the engine power (mean effective pressure), then examine mechanical methods of supercharging. We follow with turbocharging, first briefly describing an exhaust turbocharger and method for calculating the mean exhaust temperature. A simple turbocharging system is examined followed by a detailed thermodynamic study of some ideal turbocharging systems. This study will form the basis for establishing matching criteria for the efficient conversion of the exhaust energy. Constant pressure and pulse charging are examined and the influence of the exhaust system design assessed. A method is described for matching both the thermodynamic and flow characteristics of turbines, compressors and turbines. A brief note on two-stage turbocharging is included. The chapter concludes with some turbocharged engine performance characteristics.

10.1 RELATIONSHIP BETWEEN TRAPPED CONDITIONS AND MEAN EFFECTIVE PRESSURE

If Q_{VS} is the energy available per standard volume of fuel/air mixture under stoichiometric conditions, then for an equivalence ϕ less than unity the maximum available energy per standard volume (density ρ_A) of fuel/air mixture is approximately ϕQ_{VS}. If the engine cylinder volume V_t at the commencement of the compression stroke contains pure air density ρ_t, then the maximum available energy when the fuel is injected will be

$$\phi \left(\frac{\rho_t}{\rho_A}\right) V_t Q_{VS}.$$

If residuals are present in the cylinder, then this expression is modified by the scavenge efficiency η_{SC} to give

$$E_{max} = \eta_{SC}\,\phi\left(\frac{\rho_t}{\rho_A}\right)V_t Q_{VS}. \tag{10.1}$$

Now the power P developed by the engine will be

$$P = p_m V_S N = \eta_{TH} E_{max} N, \tag{10.2}$$

where p_m is the mean effective pressure, V_S is the swept volume and N the number of effective cycles per second. The thermal efficiency η_{TH} may be brake or indicated depending on whether the power P is at the brake or in the cylinder.

Combining (10.1) and (10.2) we have the mean effective pressure p_m given by

$$p_m = \eta_{TH}\eta_{SC}\phi\left(\frac{\rho_t}{\rho_A}\right)\left(\frac{V_t}{V_S}\right)Q_{VS}. \tag{10.3}$$

It will be seen from this expression that in order to increase the mean effective pressure we must increase the trapped density ρ_t. The methods whereby we attempt to do this are called supercharging. There are basically three methods. The first is to use the gas dynamic, i.e. wave action, effects in the intake and exhaust manifolds. Two methods may be used, the most well known is manifold tuning, the least well known is by a device called a comprex(1) in which the effect of wave action between hot and cold interfaces is used to raise the air pressure. The second basic method of supercharging is to provide a separate pump to supply air to the engine. The air pump may be driven from the crankshaft or connecting rod or by a separate electric motor. This method is normally called mechanical supercharging. The third basic method is to drive the compressor by utilizing some of the energy available in the exhaust gases. In this case the gas is expanded in a gas turbine which drives the compressor. This method is called exhaust turbocharging or simply turbocharging. In some engines, combinations of two or all three of these basic methods may be employed.

When the air is supplied from an external source and compressed before entering the engine, the air temperature is raised, thus lowering the effective density ρ_t. To obtain maximum utilization of the supercharged air it is usual to cool the air in some form of cooler, usually a water cooler. Thus a supercharging system will contain a compressor and an air cooler.

The flow characteristics of the air compressor and the engine must be matched for effective supercharging. If an engine-driven compressor is fitted, the compressor speed must be directly related to the engine speed in such a manner that the scavenge ratio in a two-stroke engine or the volumetric efficiency in a four-stroke engine is constant over the speed range. For exhaust turbocharging systems there are two separate sets of criteria which must be met if the turbocharger is to be matched over the whole speed/power range, these correspond to the energy recovered from the exhaust gases and the flow characteristics of the turbine and compressor.

We shall first examine mechanical supercharging and then turbocharging.

10.2 MECHANICAL SUPERCHARGING

Six types of mechanical superchargers may be used, these are:

(1) Centrifugal compressors.
(2) Axial compressors.
(3) Rotating vane compressors.
(4) Screw compressors.
(5) Rotary displacement—Roots blower.
(6) Reciprocating displacement—pump.

The same principles apply to all types. We specifically refer to the Roots blower. This is normally driven from the engine crankshaft through a gear or belt system (Fig. 10.1). The blower runs with a fixed speed ratio referred to the crankshaft. A typical pressure-mass flow characteristic is shown in Fig. 10.2(a). The full lines correspond to constant compressor speed and the broken lines to adiabatic efficiency. The compressor provides air to the cylinder under pressure. In order to match the compressor characteristics to the engine, we must obtain the engine characteristics in the same form as the compressor. We shall illustrate the procedure and the matching by a simple example.

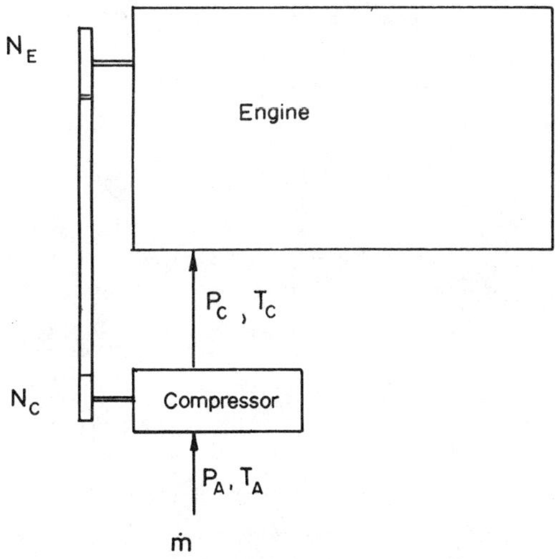

FIG. 10.1. Mechanical supercharging.

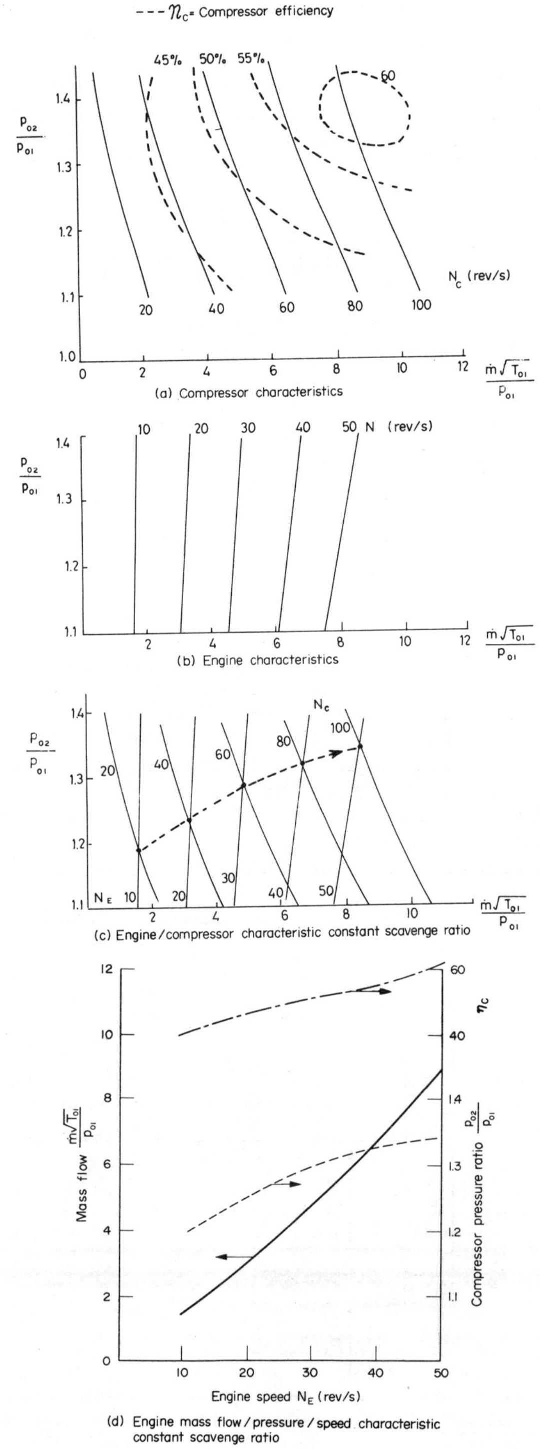

FIG. 10.2. Matching rotary displacement compressor to engine.

Consider a two-stroke engine which is to be supplied with air at pressure p_{02} with scavenge ratio λ. If the delivery temperature to the engine is T_{02}, the density ρ_{02}, V the trapped volume and N_E the engine speed, then the air supply rate $\dot{m}$ will be

$$\dot{m} = \rho_{02}\ \lambda\ VN_E. \tag{10.4}$$

Now
$$\rho_{02} = \frac{p_{02}}{RT_{02}},$$

and expression (10.4) can be rearranged to give

$$\frac{\dot{m}\sqrt{T_{01}}}{p_{01}} = \left(\frac{p_{02}}{p_{01}}\right)\left(\frac{T_{01}}{T_{02}}\right)\frac{1}{\sqrt{T_{01}}}\,\frac{1}{R}\,\lambda VN_E \tag{10.5}$$

when p_{01}, T_{01} are the inlet pressure and temperature to compressor.

Now for the compressor,

$$\frac{T_{02}}{T_{01}} = \left(\frac{p_{02}}{p_{01}}\right)^{\frac{n-1}{n}}$$

where n is the polytropic index; it follows from (10.5) that

$$\frac{\dot{m}\sqrt{T_{01}}}{p_{01}} = \left(\frac{p_{02}}{p_{01}}\right)^{\frac{1}{n}}\frac{1}{\sqrt{T_{01}}}\,\frac{1}{R}\,\lambda\ VN_E. \tag{10.6}$$

Since T_{01}, R, V are constant we can obtain the engine characteristics as shown in Fig. 10.2(b) for fixed values of scavenge ratio and various speeds N_E.

In this example we will assume that the ratio of the compressor speed N_C to the engine speed N_E is 2 to 1. We can then superimpose the two characteristics, Fig. 10.2(a) and (b), and locate the boost pressure ratio (p_{02}/p_{01}), mass flow characteristics $\left((\dot{m}\sqrt{T_{01}})/(p_{01})\right)$, for the system from the intersection of the corresponding speed lines such that $N_C = 2N_E$ (Fig. 10.2(c)). The resulting mass flow pressure ratio speed characteristics for the engine compressor system are shown in Fig. 10.2(d).

The engine characteristics given in Fig. 10.2(b) will depend on the port area designs and are used as the basis for the calculation for the port area as described in Chapter 7. It may not be possible to obtain these characteristics, and some form of compromise will be necessary.

By replacing the scavenge ratio by the volumetric efficiency, a four-stroke cycle engine match can be made in the same manner as a two-stroke engine.

The power required to drive the compressor is obtained in the usual manner. Using positive numbers

$$\dot{W}_C = \dot{m}\Delta h = \dot{m}C_p\Delta T$$

and $$\Delta T = T_{02}-T_{01}.$$

If we define the compressor efficiency as

$$\eta_C = \frac{T_{02'}-T_{01}}{T_{02}-T_{01}} = \frac{\text{isentropic temperature rise}}{\text{actual temperature rise}},$$

where $$T_{02'} = T_{01}\left(\frac{p_{02}}{p_{01}}\right)^{\frac{\kappa-1}{\kappa}},$$

then the compressor power will be

$$\dot{W}_C = \frac{\dot{m}C_pT}{\eta_C}\left\{\left(\frac{p_{02}}{p_{01}}\right)^{\frac{\kappa-1}{\kappa}} - 1\right\}. \qquad (10.7)$$

The engine power to drive the compressor $\dot{W}_{CE}$ will be greater than $\dot{W}_C$ due to mechanical losses in the drive. If we let η_M be the drive mechanical efficiency, then the engine power to drive the compressor will be

$$\dot{W}_{CE} = \frac{\dot{m}C_pT}{\eta_M\eta_C}\left\{\left(\frac{p_{02}}{p_{01}}\right)^{\frac{\kappa-1}{\kappa}} - 1\right\}. \qquad (10.8)$$

The power to drive the compressor must be deducted from the indicated power; thus a mechanical supercharger does not enable the maximum advantage to be taken of supercharging. An alternative type of supercharging is normally preferred; this utilizes the exhaust turbocharger which we shall now examine.

10.3 TURBOCHARGER

A turbocharger comprises a gas turbine and a compressor. The gas turbines in small turbochargers are normally radial turbines, whilst in a large turbocharger they are always single-stage axial turbines. The compressors are always single-stage centrifugal compressors with radial impellers and vaneless diffusers. A typical small turbocharger is illustrated in Fig. 10.3 and a large turbocharger in Fig. 10.4. The characteristics of the two types of turbine are shown in Figs. 10.5(a) and 10.5(b) and Figs. 10.6(a) and 10.6(b), respectively. The characteristics of a typical centrifugal

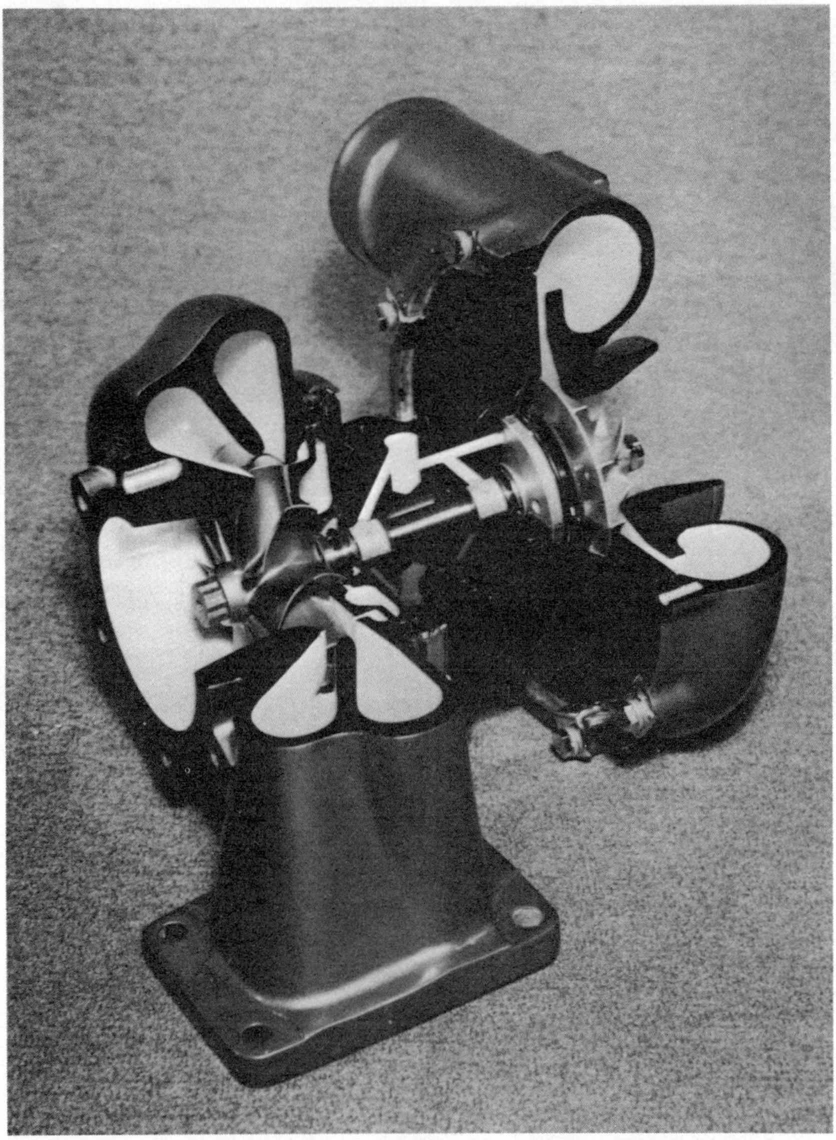

FIG. 10.3. A typical small turbocharger.

compressor are shown in Fig. 10.7. The basic difference between the two types of turbine are the speed dependent flow characteristics of radial turbines. The turbine total to static efficiency (η_{TS}) will vary with blade speed ratio (u/c_{TS}), with turbine rotor speed ($N/\sqrt{T}$) and with admission width (see Figs. 10.5(b) to 10.6(b)). In an exhaust system on an engine the turbine is presented with pulses of fluctuating pressure and temperature and this produces variations of blade speed ratio and total-to-static efficiency with time; a typical example is shown in Fig. 10.8. Unless otherwise stated we shall refer to an average turbine efficiency η_T which will allow for the fluctuations. This is given by the expression

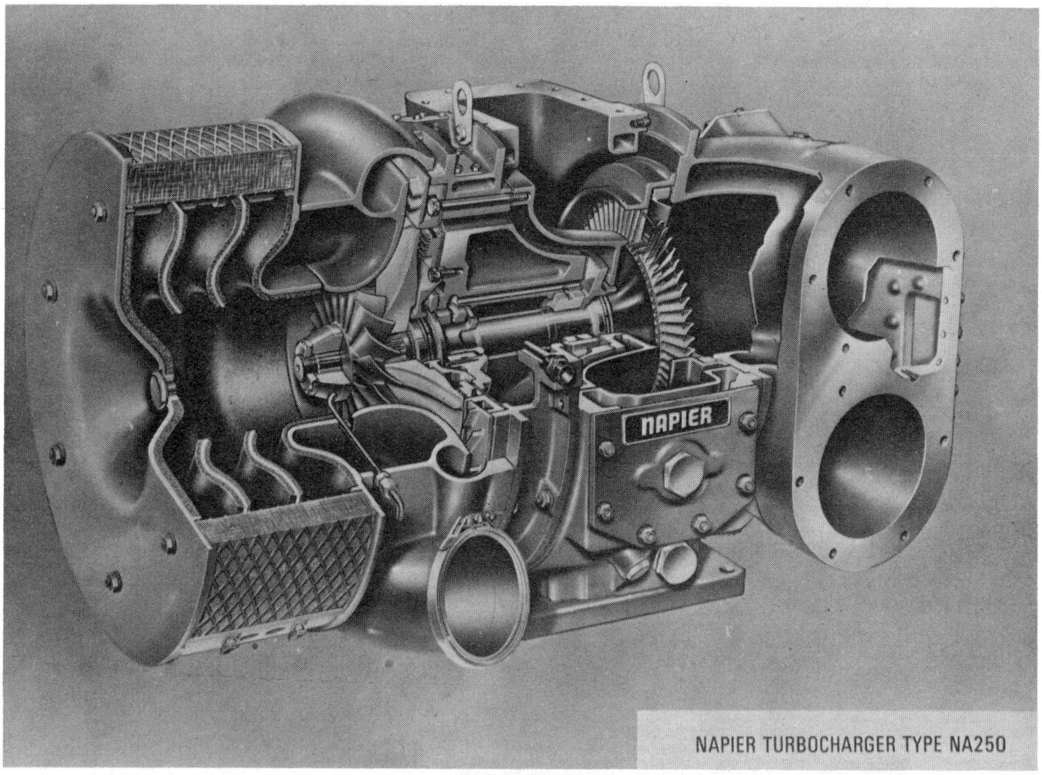

FIG. 10.4. A large turbocharger.

$$\eta_T = \frac{\int \eta_{TS}\, \dot{m}\Delta h_{is} dt}{\int \dot{m}\Delta h_{is} dt} \tag{10.9}$$

where $\dot{m}$ is the instantaneous mass flow, Δh_{is} the instantaneous isentropic enthalpy drop and η_{TS} the instantaneous total-to-static efficiency.

The exhaust gas enters the turbine either in a single- or multi-entry casing. The latter, called partial admission entry, is normally used to avoid interference between the exhausting processes of different cylinders. Interference between cylinders depends not only on cylinder locations and firing order but also on turbine location. To avoid interference, cylinders are grouped together (Fig. 10.9). The locations of the turbine have been studied in some detail by Benson and Wild[2,3] by experiment. Later we shall indicate briefly how analytical methods may be used to assist in obtaining the correct location.

The compressor air is normally delivered into a common manifold, but there are complex systems in which the compressor delivers to a displacement compressor. We shall discuss the reasons for this arrangement later.

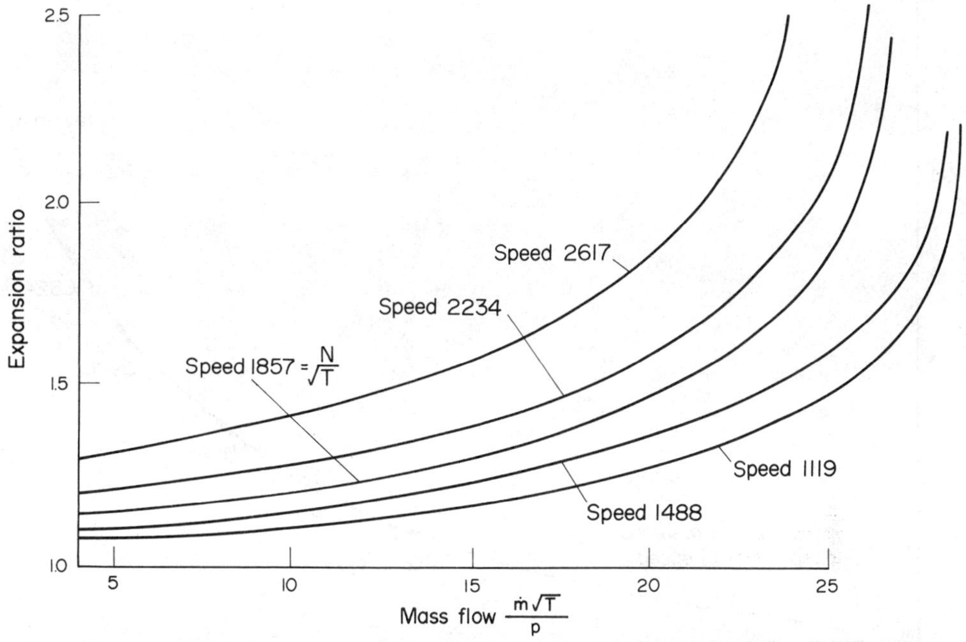

FIG. 10.5(a). Radial turbine mass flow characteristics. ṁ, flow rate; T, inlet temperature; p, initial pressure; N, rotor rev/time.

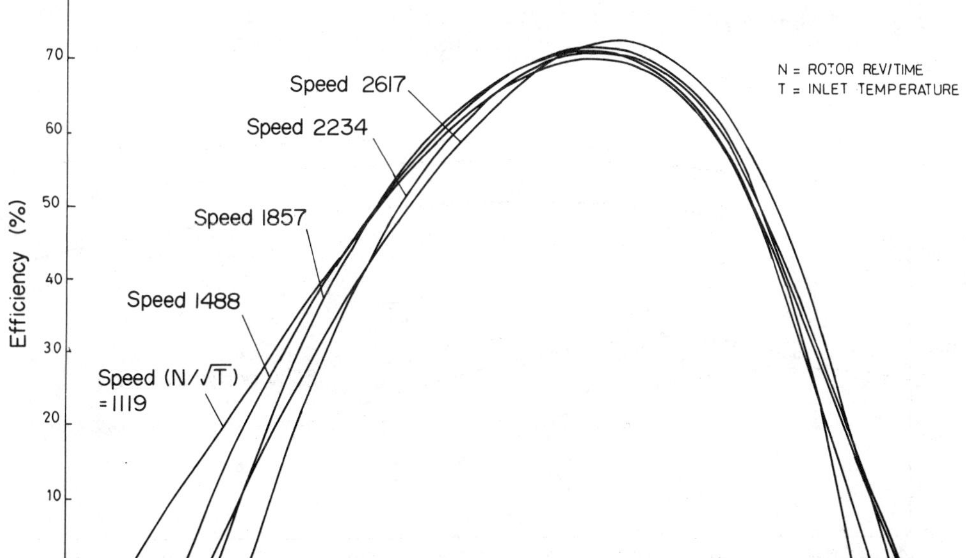

FIG. 10.5(b). Radial turbine efficiency. N, rotor rev/time; T, inlet temperature.

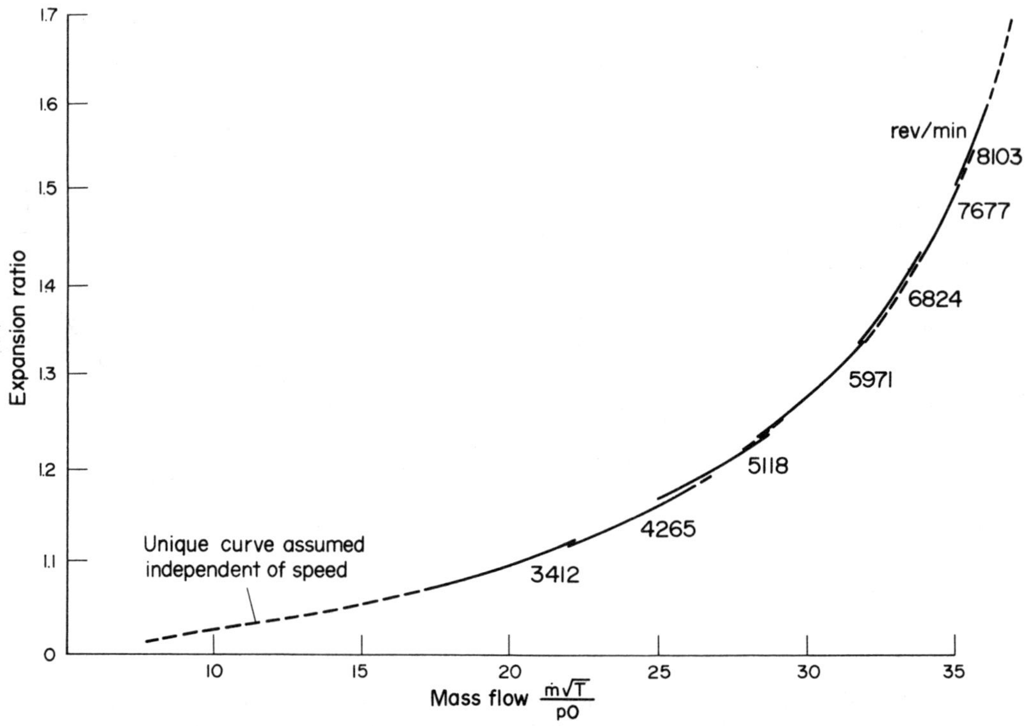

FIG. 10.6(a). Axial turbine characteristics.

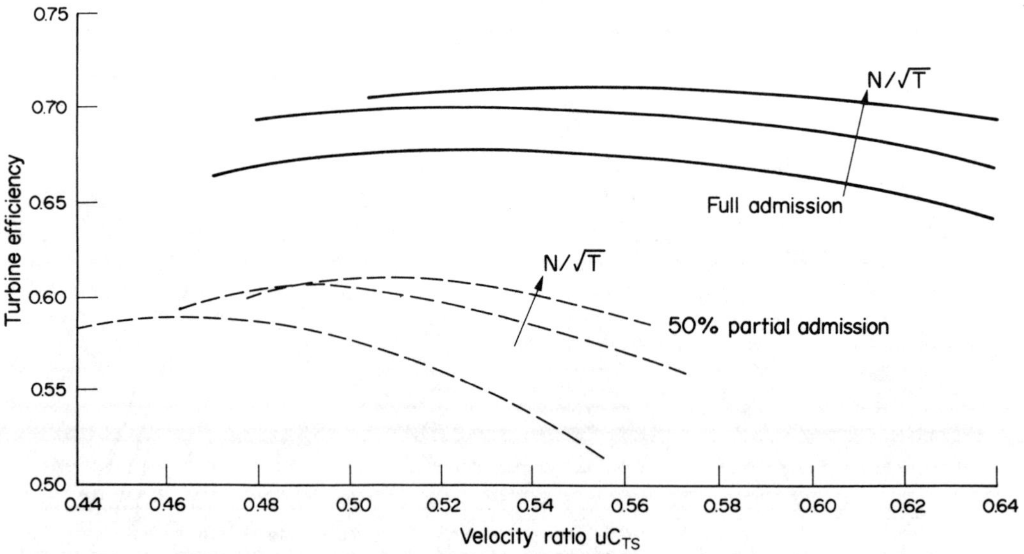

FIG. 10.6(b). Axial turbine efficiency.

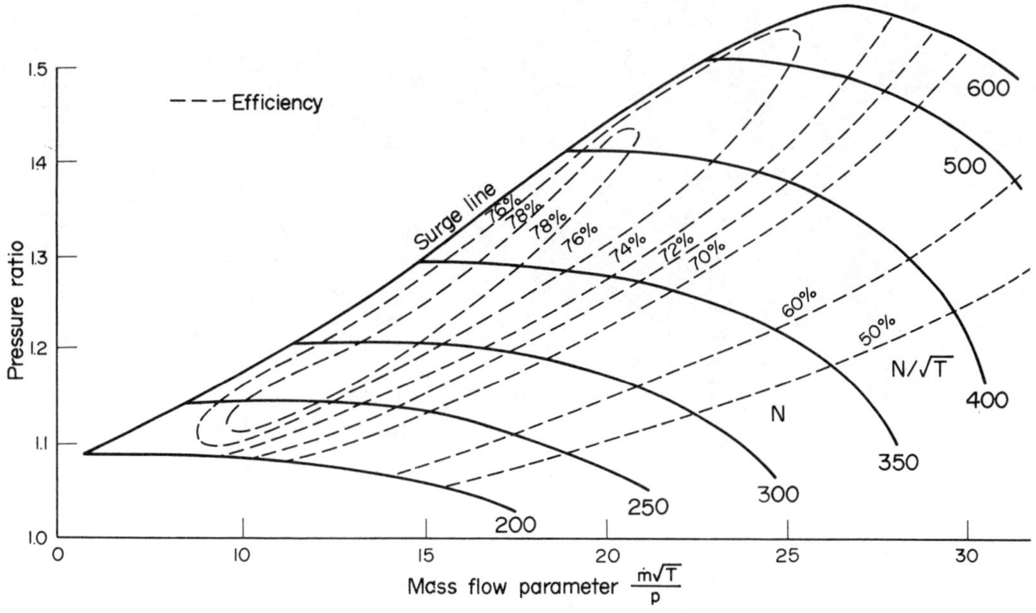

FIG. 10.7. Compressor characteristics. $\dot{m}$, mass flow rate; T, inlet temperature; p, inlet pressure; N, rev/time.

In order to economize in manufacture, turbochargers are made in a limited number of frame sizes and matched to the engine by adjusting nozzle exit dimensions (width and angle) and compressor diffuser width. Sometimes rotors have cropped blades or impellers, but this is not normal. The matching of the turbocharger to the engine can now be carried out very closely by thermodynamic and gas dynamic analysis, but because of vagaries in manufacture the final match is normally carried out on the test bed in the prototype engine. In the following sections the basic thermodynamics of the matching process are discussed.

10.4 MEAN EXHAUST TEMPERATURE

The energy required to drive the turbine is provided by the exhaust gases. The exhaust temperature and pressure will fluctuate with time, hence so will the available enthalpy drop across the turbine. We can, however, derive a mean exhaust temperature which can be used in the thermodynamic study of the turbocharger system. The engine may be represented by a box (Fig. 10.10) in which air $\dot{m}_a$ and fuel $\dot{m}_f$ are supplied and gas leaves the box before entering the turbine with enthalpy h_{03}.

The first law for the control system enclosed by the engine is

$$\dot{Q}-\dot{W}_S = \dot{m}_{out}h_{out} - \dot{m}_{in}h_{in}.$$

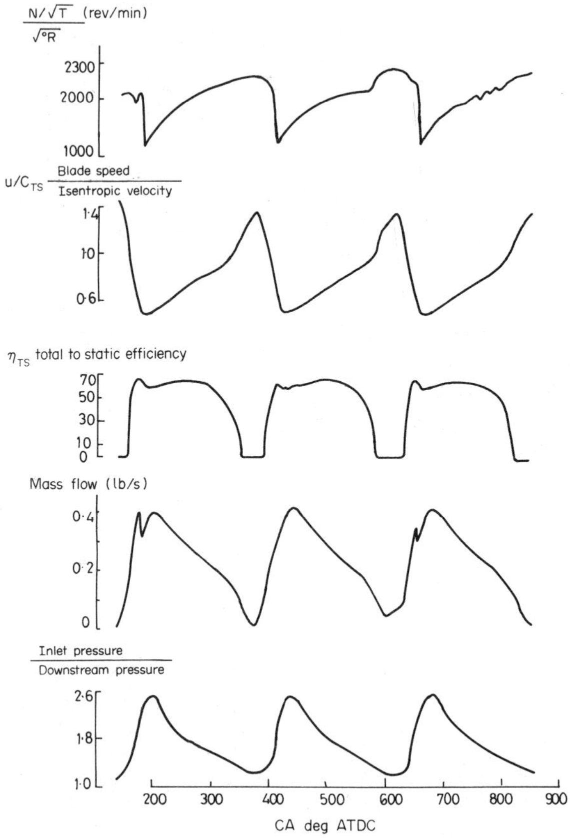

FIG. 10.8. Transient characteristics of turbocharger turbine.

The mass flow out $\dot{m}_{out}$ is the sum of the air flow $\dot{m}_a$ and the fuel flow $\dot{m}_f$.

$$\dot{m}_{out} = \dot{m}_a + \dot{m}_f.$$

When we consider the enthalpy, this is comprised of two parts—the enthalpy at absolute zero plus a temperature dependent term, i.e.

$$h = h_0 + C_p T,$$

where C_p is the mean specific heat from 0 to T.

For the products ($C_p = C_{pE}$).

$$\dot{m}_{out} h_{out} = (\dot{m}_a + \dot{m}_f)(h_{0P} + C_{pE} T_{03}).$$

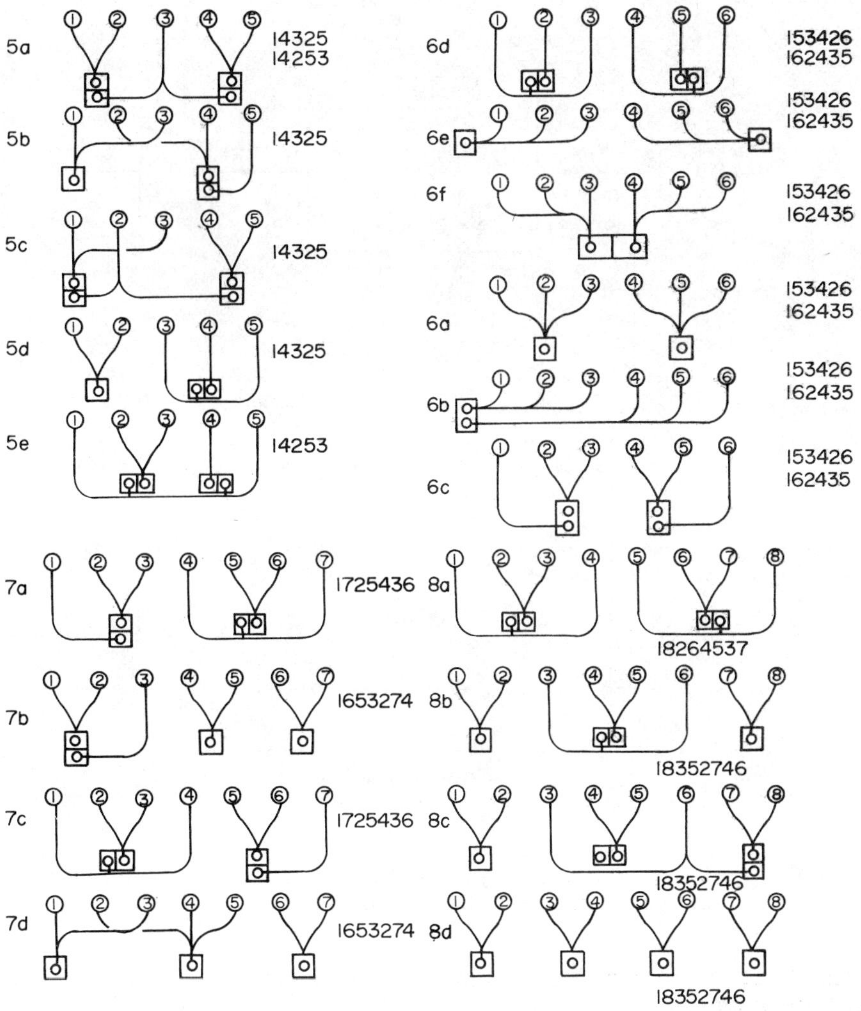

FIG. 10.9. Exhaust pipe configurations for supercharging systems. Encircled numbers (e.g. ⑤) indicates number of cylinders. Sequence of numbers (e.g. 14325) indicates firing order.

For the reactants ($C_p = C_{pa}$),

$$\dot{m}_{in} h_{in} = (\dot{m}_a + \dot{m}_f) h_{0R} + C_{pa} T_{02} + C_{pf} T_f .$$

We can neglect the term $C_{pf} T_f$ for the fuel.

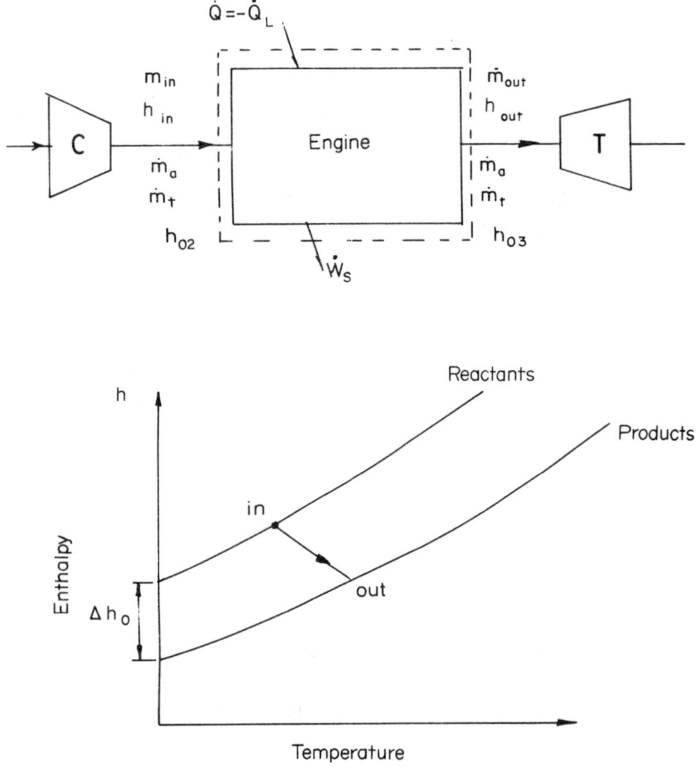

FIG. 10.10. Control system for calculating mean exhaust temperatures. C, compressor; T, turbine.

The heat transfer $\dot{Q}$ is equal to the heat loss $-\dot{Q}_L$, where $\dot{Q}_L$ is a positive number.

Hence the first law for the system becomes

$$-\dot{Q}_L-\dot{W}_S = (\dot{m}_a+\dot{m}_f)C_{pE}T_{03} - \dot{m}_aC_{pa}T_{02} + (\dot{m}_a+\dot{m}_f)(h_{0P}-h_{0R}).$$

The term $(h_{0P}-h_{0R})$ is the specific enthalpy of reaction (or heat of reaction) at the absolute zero and is given the symbol Δh_0 and $(\dot{m}_a+\dot{m}_f)(h_{0P}-h_{0R}) = \Delta H_0$, the total heat of reaction.

Hence

$$-\Delta H_0-\dot{Q}_L-\dot{W}_S = (\dot{m}_a+\dot{m}_f)C_{pE}T_{03} - \dot{m}_aC_{pa}T_{02}.$$

ΔH_0 is negative for an exothermic reaction; if $-Q_f$ is the heat of reaction per mass of fuel, where Q_f is a positive number, then

$$\Delta H_0 = -\dot{m}_f Q_f$$

and the first law becomes

$$\dot{m}_f Q_f - \dot{Q}_L - \dot{W}_S = (\dot{m}_a + \dot{m}_f) C_{pE} T_{03} - \dot{m}_a C_{pa} T_{02}. \quad (10.10)$$

The thermal efficiency of the engine η_{TH} is

$$\eta_{TH} = \frac{\dot{W}_S}{\dot{m}_f Q_f}.$$

The percentage heat loss is

$$\zeta_L = \frac{\dot{Q}_L}{\dot{m}_f Q_f}.$$

If we let
$$\zeta_E = 1 - \zeta_L - \eta_{TH}$$

and divide (10.10) by $(\dot{m}_a + \dot{m}_f) C_{pE}$, then

$$\frac{\dot{m}_f}{\dot{m}_a + \dot{m}_f} \frac{Q_f}{C_{pE}} \left(1 - \zeta_L - \eta_{TH}\right) = T_{03} - \frac{\dot{m}_a}{\dot{m}_a + \dot{m}_f} \left(\frac{C_{pa}}{C_{pE}}\right) T_{02}$$

or

$$\frac{1}{\left((\dot{m}_a/\dot{m}_f) + 1\right)} \frac{Q_f}{C_{pE}} \zeta_E = T_{03} - \left(\frac{\dot{m}_a/\dot{m}_f}{(\dot{m}_a/\dot{m}_f) + 1}\right) \left(\frac{C_{pa}}{C_{pE}}\right) T_{02}, \quad (10.11)$$

where $\dot{m}_a/\dot{m}_f$ is the overall air/fuel ratio A/F and C_{pa}/C_{pE} is the ratio of the mean specific heats.

The term

$$\left(\frac{\dot{m}_a/\dot{m}_f}{(\dot{m}_a/\dot{m}_f) + 1}\right) \left(\frac{C_{pa}}{C_{pE}}\right)$$

varies from about 0.92 to 1.00 over the normal air/fuel ratio temperature range, thus (10.11) can be approximated to

$$\left(\frac{1}{1 + (\dot{m}_a/\dot{m}_f)}\right) \left(\frac{Q_f}{C_{pE}}\right) \zeta_E = T_{03} - T_{02} = \Delta T. \quad (10.12)$$

The group $(Q_f/C_{pE})\zeta_E$ is related to the heat loss and thermal efficiency. The thermal efficiency will vary with the air/fuel ratio and the heat loss with air/fuel ratio and speed.

If we set

$$K = \frac{Q_f}{C_{pE}}\zeta_E,$$

then

$$K = K(\dot{m}_a/\dot{m}_f,\ N_E) \text{ for a given engine.} \quad (10.13)$$

Thus

$$\Delta T = \frac{K}{1 + (\dot{m}_a/\dot{m}_f)} \quad (10.14)$$

or

$$T_{03} = T_{02} + \frac{K}{1 + (\dot{m}_a/\dot{m}_f)}. \quad (10.15)$$

In Fig. 10.11 K is plotted against $\dot{m}_a/\dot{m}_f$ for a typical engine. It will be observed that over a wide range of air/fuel ratio it is almost constant but then it falls with low air/fuel ratio. This is due to poor combustion and some of the fuel passing directly through the engine without burning at all.

If we examine the expression for ζ_E,

$$\zeta_E = 1 - \zeta_L - \eta_{TH}. \quad (10.16)$$

Then, since $\eta_{TH} + \zeta_L + \zeta_E = 1.0$, ζ_E is a measure of the energy available in the exhaust gas.

The temperature T_{03} is not the instantaneous exhaust temperature ahead of the turbine but an equivalent mean exhaust temperature which can be used in a similar manner to the average turbine efficiency given in expression (10.9).

10.5 SIMPLE TURBOCHARGING SYSTEM

A simple turbocharger system is shown in Fig. 10.12. Air from the atmosphere is compressed in compressor C from pressure p_{01} to p_{02} and temperature T_{01} to T_{02}'. The air is then cooled in the intercooler and passes to the engine. Exhaust gas leaves the cylinders and enters a large manifold or receiver at a mean pressure p_{03} and temperature T_{03}. It is then expanded in the turbine to the back pressure p_{04}. The turbine drives the compressor. The thermodynamic analysis uses the first law of thermodynamics for the control systems enclosing the turbine and the compressor separately. We shall use the positive values for power.

When the turbine work equals the compressor work allowing for mechanical losses, the system is in equilibrium. If we let the

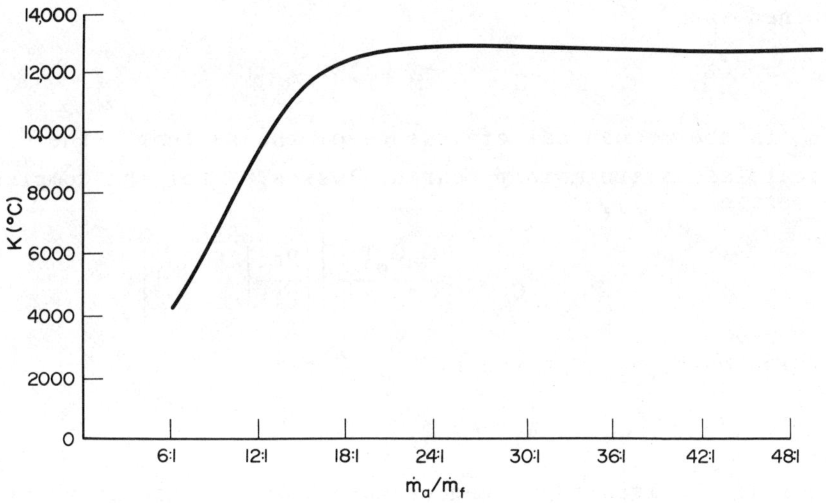

FIG. 10.11. Engine temperature rise factor K.

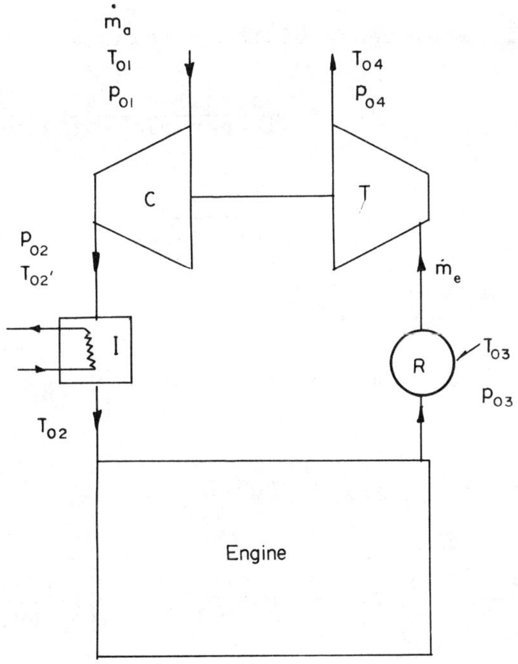

FIG. 10.12. Simple turbocharger system. C, compressor; T, turbine; I, intercooler; R, receiver.

turbine power be $\dot{W}_T$ and the compressor power be $\dot{W}_C$, then equilibrium is obtained when

$$\dot{W}_C = \eta_M \dot{W}_T,$$

where η_M is the mechanical efficiency of the system. The mechanical losses are mainly bearing losses. For the compressor $\dot{W}_C$ is, as before (10.7),

$$\dot{W}_C = \frac{\dot{m}_a C_p T}{\eta_C}\left[\left(\frac{p_{02}}{p_{01}}\right)^{\frac{\kappa-1}{\kappa}} - 1\right].$$

The turbine power $\dot{W}_T$ is given by

$$\dot{W}_T = \dot{m}_e C_{pg}(T_{03}-T_{04}).$$

If we define the isentropic temperature $T_{04'}$, due to expansion from p_{03} to p_{04} from temperature T_{03} as

$$T_{04'} = T_{03}\left(\frac{p_{04}}{p_{03}}\right)^{\frac{\kappa_g-1}{\kappa_g}}$$

and the isentropic efficiency of the turbine as

$$\eta_T = \frac{\text{actual enthalpy drop}}{\text{isentropic enthalpy drop}}$$

$$\frac{T_{03}-T_{04}}{T_{03}-T_{04'}},$$

then the turbine work is

$$\dot{W}_T = \dot{m}_e \eta_T C_{pg} T_{03}\left[1-\left(\frac{p_{04}}{p_{03}}\right)^{\frac{\kappa_g-1}{\kappa_g}}\right]. \quad (10.17)$$

At equilibrium, when $\dot{W}_C = \eta_M \dot{W}_T$,

$$\frac{\dot{m}_a C_p T}{\eta_C}\left[\left(\frac{p_{02}}{p_{01}}\right)^{\frac{\kappa-1}{\kappa}} - 1\right] = \eta_T \eta_M \dot{m}_e C_{pg} T_{03}\left[1-\left(\frac{p_{04}}{p_{03}}\right)^{\frac{\kappa_g-1}{\kappa_g}}\right]$$

or
$$\eta_M \eta_T \eta_C = \left(\frac{\dot{m}_a}{\dot{m}_e}\right)\left(\frac{C_p}{C_{pg}}\right)\left(\frac{T_{01}}{T_{03}}\right)\frac{\left\{\left(\frac{p_{02}}{p_{01}}\right)^{\frac{\kappa-1}{\kappa}} - 1\right\}}{\left\{1 - \left(\frac{p_{04}}{p_{03}}\right)^{\frac{\kappa_g-1}{\kappa}}\right\}} \tag{10.18}$$

The term $\eta_M\eta_T\eta_C$ is called the overall turbocharger efficiency η_{TC}.

If we let $\dot{m}_a = \dot{m}_e$, $C_p = C_{pg}$, $\kappa_g = \kappa$, then

$$\eta_{TC} = \frac{T_{01}}{T_{03}}\frac{\left\{\left(\frac{p_{02}}{p_{01}}\right)^{\frac{\kappa-1}{\kappa}} - 1\right\}}{\left\{1 - \left(\frac{p_{04}}{p_{03}}\right)^{\frac{\kappa-1}{\kappa}}\right\}} \tag{10.19}$$

$$\frac{p_{02}}{p_{01}} = \left\{1 + \eta_{TC}\left(\frac{T_{03}}{T_{01}}\right)\left\{1 - \left(\frac{p_{04}}{p_{03}}\right)^{\frac{\kappa-1}{\kappa}}\right\}\right\}^{\frac{\kappa}{\kappa-1}} . \tag{10.20}$$

If we examine this expression we can ascertain some of the more important physical aspects of turbocharging. Firstly, we observe that the higher the turbocharger efficiency (η_{TC}) the greater the boost (p_{02}/p_{01}); similarly, the higher the exhaust temperature T_{03} and the greater the expansion ratio across the turbine (p_{03}/p_{04}), so the greater the boost. For a given engine the exhaust temperature is dependent on the brake thermal efficiency and the heat losses. The higher the brake thermal efficiency so the lower the exhaust temperature; the lower the heat losses, the higher the exhaust temperature. The thermal load, i.e. the stress due to the engine temperature, depends on the exhaust temperature, so that it is desirable not to exceed a set value. Thus raising the exhaust temperature to achieve high boost is not to be advised. If the expansion ratio across the turbine is increased, then more work can be done and the boost pressure raised. However, there is a limit to the exhaust expansion ratio (p_{03}/p_{04}); this should not exceed the boost pressure by too high a margin otherwise the back pressure will cause additional pumping work in a four-stroke engine or retard the scavenge process in a two-stroke engine—indeed, in the two-stroke engine it should always be lower than the boost pressure ratio. Thus one is finally drawn to the main parameter—the turbocharger efficiency. If this is high, so is the boost pressure ratio; if it is low, so is the boost pressure ratio.

Now in the simple system shown in Fig. 10.12 the pressure and the temperature in the receiver are uniform and constant. In practice this is not so. However, the larger the receiver the more likely these conditions will be met with. On the other hand, if we reduce the size of the receiver both the pressures and temperatures in the receiver will fluctuate with time, and if the turbine nozzles are small compared with the total exhaust valve areas, then the pressures in the receiver might be very high and would certainly fluctuate according to the exhaust valve opening rate. In these cases the expansion ratio across the turbine will fluctuate and will be on average larger than that due to the mean pressure p_{03}, hence more work can be obtained from the turbine. In the ideal case, if we remove the receiver and attach the turbine directly to the exhaust ports of the engine and use the ports as nozzles, we could recover the maximum available energy in the exhaust gas. We will now consider such a system. A diagrammatic agreement is shown in Fig. 10.13. The turbine is close to the exhaust ports with an extremely small exhaust pipe.

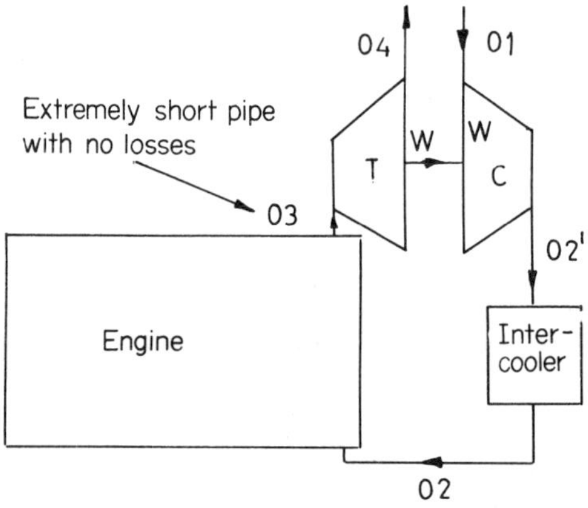

FIG. 10.13. Ideal turbocharging system.
T, turbine; C, compressor.

10.6 IDEAL TURBOCHARGING SYSTEM

We shall examine the ideal system for the three parts of the engine gas exchange process in which the cylinder is exposed to the turbine, namely:

(1) during exhaust blowdown;

(2) during the exhaust stroke in a four-stroke engine;

(3) when both air and exhaust valves (or ports) are open.

We shall assume the following:

(1) The gas exchange process can be represented by the pressure diagrams shown in Fig. 10.14.

(2) The cylinder pressure varies linearly with volume during exhaust blowdown.

(3) During the exhaust stroke and when both air and exhaust valves (or ports) are open the cylinder pressure is constant.

(4) During exhaust blowdown the exhaust pressure is <u>not</u> constant.

(5) During the exhaust stroke and when both air and exhaust valves (or ports) are open the exhaust pressure is constant.

(6) The boost pressure is constant.

(7) The scavenge process is adiabatic.

(8) There are no losses in the expansion of the gas from the exhaust pressure through the turbine.

(9) The ideal enthalpy/entropy diagrams are as shown in Fig. 10.15.

We shall define the pressure ratios:

$\alpha = \frac{p}{p_{02}}$, p_{03} = exhaust pressure,

$\beta = \frac{p_{03}}{p}$ p_{02} = boost pressure,

p = cylinder pressure, p_R = cylinder release pressure.

The general first law for the control volume for a time dt is

$$dQ - dW_S - p\,dV = d(me)_{Cyl} + dm_{out}h_{out} - dm_{in}h_{in}. \tag{10.21}$$

The control surface will enclose the engine cylinders in parts of the analysis, in which case $dW_S = 0$, and in other parts of the analysis the control surface will enclose both the engine cylinders and turbine. In the latter case $dW_S = dW_T$. The cylinder internal energy is $me = pV/\kappa-1$, whence the general first law becomes

$$dQ - dW_T - p\,dV = \frac{1}{\kappa-1}(p\,dV + V\,dp) + dm_{out}h_{out} - dm_{in}h_{in} \tag{10.22}$$

or

$$dQ - dW_T - p\ dV = \frac{1}{\kappa-1}\{d(pV)\} + dm_{out}h_{out} - dm_{in}h_{in}. \tag{10.23}$$

We shall first examine the two-stroke engine (Fig. 10.14(a)) and then the four-stroke engine (Fig. 10.14(b)).

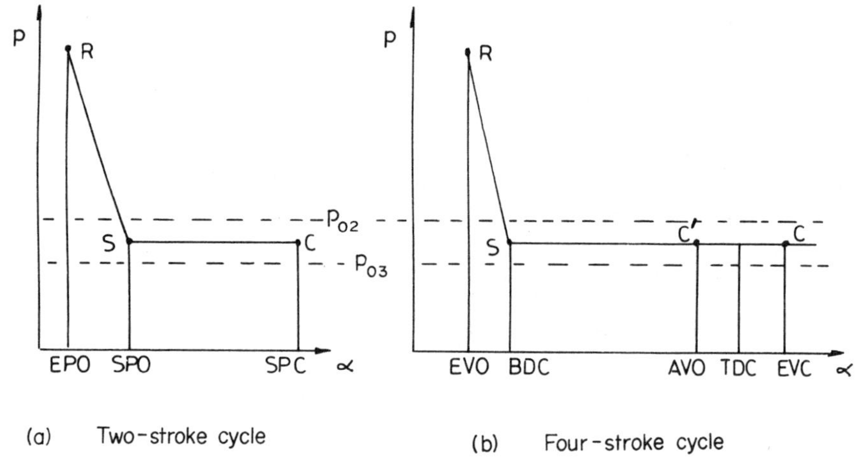

FIG. 10.14. Ideal pressure diagrams.

10.6.1 Two-stroke Engine

We shall examine the process in two parts: the exhaust blowdown and the scavenge period.

Exhaust Blowdown Period (R-S)

Applying the first law (10.23) for an adiabatic process, $dQ = 0$, to the engine plus turbine as one control volume, with $dm_{in} = 0$.

Equation (10.23) becomes

$$-dW_T - p\ dV = \frac{1}{\kappa-1}\ d(pV) + dm_4 h_{04}. \tag{10.24}$$

The mass efflux from the turbine dm_{out} equals dm_4. The specific enthalpy from the turbine is h_{04}. From the enthalpy diagram (Fig. 10.15(a)) the specific enthalpy h_{04} is a constant and is given by

$$h_{04} = h_R\left(\frac{p_{04}}{p_R}\right)^{\frac{\kappa-1}{\kappa}}. \tag{10.25}$$

Now the mass leaving the cylinder $dm_c = -dm_4$.

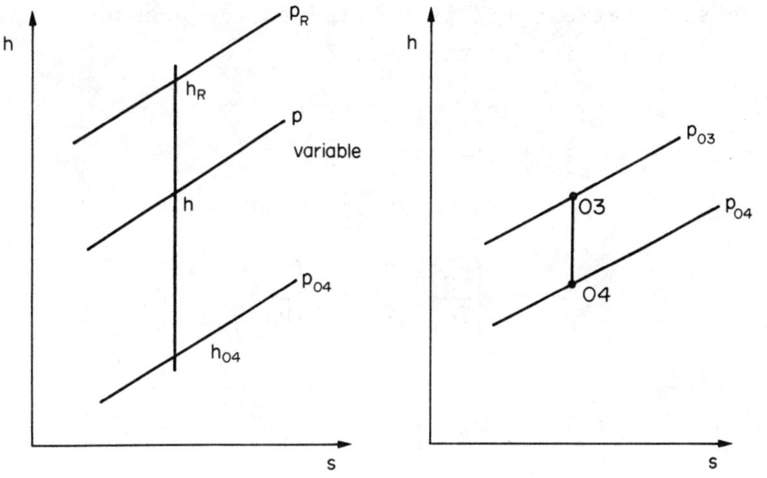

(a) Enthalpy/entropy (h/s) diagram across turbine during blowdown

(b) Enthalphy/entropy (h/s) diagram across turbine during remainder of cycle

FIG. 10.15. Enthalpy diagrams across turbine.

Substituting for dm_{04} and h_{04} in (10.24) we have

$$- dW_T = \frac{1}{\kappa-1} d(pV) - h_R\left(\frac{p_{04}}{p_R}\right)^{\frac{\kappa-1}{\kappa}} dm_c + p\ dV. \tag{10.26}$$

Integrating between EPO (R) and SPO (S) we obtain

$$- W_T = \frac{1}{\kappa-1} (p_S V_S - p_R V_R) - h_R\left(\frac{p_{04}}{p_R}\right)^{\frac{\kappa-1}{\kappa}} (m_S - m_R) + \int_R^S p\ dV. \tag{10.27}$$

For a linear variation of cylinder pressure p with volume V,

$$\int_R^S p\ dV = \tfrac{1}{2}(p_R + p_S)(V_S - V_R). \tag{10.28}$$

If we let $y = V_S/V_R$ expression (10.28) becomes

$$\int_R^S p\ dV = \tfrac{1}{2}V_R(y-1)(p_R + p_S). \tag{10.29}$$

Consider the second term on the right-hand side of (10.27):

$$m_S = \rho_S V_S, \qquad m_R = \rho_R V_R.$$

The density ratio ρ_S/ρ_R is related to the pressure ratio p_S/p_R for isentropic expansion in the cylinder

$$\frac{\rho_S}{\rho_R} = \left(\frac{p_S}{p_R}\right)^{\frac{1}{\kappa}}$$

Hence
$$\frac{m_S}{m_R} = \left(\frac{p_S}{p_R}\right)^{\frac{1}{\kappa}} \frac{V_S}{V_R} = y\left(\frac{p_S}{p_R}\right)^{\frac{1}{\kappa}}.$$

$$h_R\left(\frac{p_{04}}{p_R}\right)^{\frac{\kappa-1}{\kappa}}(m_S-m_R) = m_R h_R\left(\frac{p_{04}}{p_R}\right)^{\frac{\kappa-1}{\kappa}}\left(\left(\frac{p_S}{p_R}\right)^{\frac{1}{\kappa}} y-1\right); \qquad (10.30)$$

also $mh = \kappa pV/\kappa-1$.

The expression for turbine work (10.27) becomes

$$W_T = \frac{1}{\kappa-1}\left(p_R V_R - p_S V_S\right) + p_R V_R\left(\frac{p_{04}}{p_R}\right)^{\frac{\kappa-1}{\kappa}}$$
$$\left(\left(\frac{p_S}{p_R}\right)^{\frac{1}{\kappa}} y-1\right)\left(\frac{\kappa}{\kappa-1}\right) - \tfrac{1}{2}V_R(y-1)\left(p_R+p_S\right). \qquad (10.31)$$

It is convenient to place the term $p_R V_R$ outside all the brackets. Then (10.31) becomes

$$W_T = \frac{p_R V_R}{\kappa-1}\left(1 - \frac{p_S}{p_R}\,y\right) + \frac{\kappa}{\kappa-1}\,p_R V_R\left(\frac{p_{04}}{p_R}\right)^{\frac{\kappa-1}{\kappa}}$$
$$\left(y\left(\frac{p_S}{p_R}\right)^{\frac{1}{\kappa}} - 1\right) - \frac{p_R V_R}{2}\,(y-1)\left(1 + \frac{p_S}{p_R}\right).$$

Let
$$x = \frac{p_S}{p_R}.$$

Divide by $p_{04}V_R$ and we have

$$\left(\frac{W_T}{p_{04}V_R}\right)_{RS} = \frac{\left(\frac{p_R}{p_{04}}\right)}{\kappa-1}\left((1-xy)-\left(\frac{\kappa-1}{2}\right)(y-1)(1+x)+\kappa\left(\frac{p_R}{p_{04}}\right)^{\frac{1-\kappa}{\kappa}}\left[yx^{\frac{1}{\kappa}}-1\right]\right), \tag{10.32}$$

noting $x = \frac{p_S}{p_R}$ and $y = \frac{V_S}{V_R}$.

This expression gives the maximum available energy to drive the turbine during the blowdown period.

Scavenge Period (S-C)

For this period we take two control volumes. The first control volume encloses the engine cylinders and the turbine; the second volume encloses the engine cylinders only. We then have the following two expressions for the first law.

Engine and turbine:

$$- dW_T - p\,dV = \frac{1}{\kappa-1}\,d(pV) + dm_4h_{04}-dm_2h_{02}. \tag{10.33}$$

Engine:

$$- p\,dV = \frac{1}{\kappa-1}\,d(pV) + dm_3h_{03}-dm_2h_{02}. \tag{10.34}$$

In this case $dm_3 = dm_4$ = exhaust flow from engine and dm_2 = air flow into engine.

Expression (10.34) becomes for constant cylinder pressure

$$dm_3h_{03} = -\frac{\kappa}{\kappa-1}\,p\,dV + dm_2h_{02}, \tag{10.35}$$

and (10.33) becomes

$$- dW_T = \frac{\kappa}{\kappa-1}\,p\,dV + dm_4h_{04}-dm_2h_{02}. \tag{10.36}$$

Now

$$\frac{h_{04}}{h_{03}} = \left(\frac{p_{04}}{p_{03}}\right)^{\frac{\kappa-1}{\kappa}} = \text{constant}.$$

Then

$$dm_4h_{04} = dm\,\frac{h_{04}}{h_{03}}\,h_{03} = dm_3h_{03}\left(\frac{p_{04}}{p_{03}}\right)^{\frac{\kappa-1}{\kappa}}$$

since $dm_4 = dm_3$.

Now from (10.35)

$$dm_3 h_{03} = -\frac{\kappa}{\kappa-1}\, p\, dV + dm_2 h_{02}.$$

$$dm_4 h_{04} = \left(\frac{p_{04}}{p_{03}}\right)^{\frac{\kappa-1}{\kappa}} \left(-\frac{\kappa}{\kappa-1}\, p\, dV + dm_2 h_{02}\right). \qquad (10.37)$$

Substituting into (10.36) we have after simplification,

$$-dW_T = \left(1-\left(\frac{p_{04}}{p_{03}}\right)^{\frac{\kappa-1}{\kappa}}\right)\left(\frac{\kappa}{\kappa-1}\, p\, dV - dm_2 h_{02}\right).$$

Integrating between SPO (S) and EPC (C) we obtain the turbine work

$$W_T = \left(1-\left(\frac{p_{04}}{p_{03}}\right)^{\frac{\kappa-1}{\kappa}}\right)\left(\frac{\kappa}{\kappa-1}\, p_S(V_S - V_C) + m_a h_{02}\right). \qquad (10.38)$$

where m_a is the total air supplied.

Now $\alpha = \dfrac{p}{p_{02}}$ and $p_S = p$.

Therefore $p_S = \alpha p_{02}$.

If V_{02} is the volume of air supplied at p_{02}, T_{02},

then $m_a h_{02} = \dfrac{\kappa}{\kappa-1}\, p_{02} V_{02} = \dfrac{\kappa\lambda}{\kappa-1}\, p_{02} V_C$,

where $\lambda = \dfrac{V_{02}}{V_C}$ = scavenge ratio.

Let $z = \dfrac{V_C}{V_R}$

Then (10.38) becomes

$$W_T = \left(1-\left(\frac{p_{04}}{p_{03}}\right)^{\frac{\kappa-1}{\kappa}}\right)\left(\frac{\lambda\kappa}{\kappa-1}\, z\, V_R p_{02} - \frac{\kappa}{\kappa-1}\, z\alpha p_{02} V_R\left(1-\frac{y}{z}\right)\right) \qquad (10.39)$$

and dividing by $p_{04} V_R$,

$$\left(\frac{W_T}{p_{04}V_R}\right)_{SC} = \left(\frac{p_{02}}{p_{04}}\right)\left(1-\left(\frac{p_{04}}{p_{03}}\right)^{\frac{\kappa-1}{\kappa}}\right)\left(\frac{z\kappa}{\kappa-1}\right)\left(\lambda-\alpha\left(1-\frac{y}{z}\right)\right), \qquad (10.40)$$

where $z = \frac{V_C}{V_R}$, $y = \frac{V_S}{V_R}$,

$\alpha = \frac{p_S}{p_{02}}$, λ = scavenge ratio.

This expression gives the maximum available energy to drive the turbine during the scavenge period.

We now turn to the compressor work.

Compressor Work

From Fig. 10.13 the compressor work for a flow of m_{02} air per cycle is

$$W_C = m_{02}(h_{02'}-h_{01}). \qquad (10.41)$$

For minimum compressor work the compression is isentropic and

$$\frac{h_{02'}}{h_{01}} = \left(\frac{p_{02'}}{p_{01}}\right)^{\frac{\kappa-1}{\kappa}} = \left(\frac{p_{02'}}{p_{02}}\frac{p_{02}}{p_{01}}\right)^{\frac{\kappa-1}{\kappa}}. \qquad (10.42)$$

Substituting into (10.41) and rearranging we have

$$W_C = m_{02}h_{02}\left(\frac{h_{01}}{h_{02}}\right)\left(\left(\frac{p_{02'}}{p_{02}}\frac{p_{02}}{p_{01}}\right)^{\frac{\kappa-1}{\kappa}} - 1\right), \qquad (10.43)$$

noting that $z = \frac{V_C}{V_R}$ and $\lambda = \frac{V_{02}}{V_C}$.

Then $$m_{02}h_{02} = \frac{\kappa}{\kappa-1}\, p_{02}V_{02} = \frac{\kappa}{\kappa-1}\,\lambda z\, p_{02}V_R. \qquad (10.44)$$

Let $v = \frac{p_{02}}{p_{02'}}$ the pressure ratio across the intercooler,

$r = \frac{p_{02}}{p_{01}}$, the boost pressure ratio to the engine, and $$n = \frac{h_{01}}{h_{02}} = \frac{T_{01}}{T_{02}}.$$

Substituting (10.44) and v, r, n into (10.43) and rearranging, we obtain the compressor work as

$$\frac{W_C}{p_{04}V_R} = \frac{n\lambda zr\kappa}{\kappa-1}\left[\left(\frac{r}{v}\right)^{\frac{\kappa-1}{\kappa}} - 1\right]. \tag{10.45}$$

We can summarize then for the two-stroke engine. The maximum work by the turbine is equal to the maximum available energy at the turbine; this comprises two components:

(1) Blowdown work from EPO to SPO

$$A = \left(\frac{W_T}{p_{04}V_R}\right)_{RS} = \frac{\left(\frac{p_R}{p_{04}}\right)}{(\kappa-1)}\left[(1-xy) - \frac{\kappa-1}{2}(y-1)(1+x) + \kappa\left(\frac{p_R}{p_{04}}\right)^{\frac{1-\kappa}{\kappa}}\left[yx^{\frac{1}{\kappa}} - 1\right]\right]. \tag{10.32}$$

(2) Scavenge work from SPO to EPC

$$B = \left(\frac{W_T}{p_{04}V_R}\right)_{SC} = \left(\frac{p_{02}}{p_{04}}\right)\left(1-\left(\frac{p_{04}}{p_{03}}\right)^{\frac{\kappa-1}{\kappa}}\right)\left(\frac{z\kappa}{\kappa-1}\right)\left(\lambda-\alpha\left(1 - \frac{y}{z}\right)\right). \tag{10.40}$$

The <u>minimum</u> work to compress the air

$$\frac{W_C}{p_{04}V_R} = \frac{n\lambda zr\kappa}{(\kappa-1)}\left[\left(\frac{r}{v}\right)^{\frac{\kappa-1}{\kappa}} - 1\right]. \tag{10.45}$$

The various pressure, temperature and volume ratios are:

$$x = \frac{p_S}{p_R}, \quad y = \frac{V_S}{V_R}, \quad z = \frac{V_C}{V_R}$$

$$\alpha = \frac{p_S}{p_{02}}, \quad r = \frac{p_{02}}{p_{01}}, \quad v = \frac{p_{02}}{p_{02'}}$$

$$n = \frac{T_{01}}{T_{02}}.$$

The scavenge ratio λ is

$$\lambda = \frac{V_{02}}{V_C}.$$

The total maximum available energy at the turbine is

$$\left(\frac{W_T}{p_{04}V_R}\right)_{tot} = \left(\frac{W_T}{p_{04}V_R}\right)_{RS} + \left(\frac{W_T}{p_{04}V_R}\right)_{SC} = A + B. \qquad (10.46)$$

In Fig. 10.16 the maximum available energy is shown for no pressure losses across the intercooler and ports $\alpha = 1.0$, $v = 1.0$. For 100% intercooling $n = 1.0$ and the volume ratio y and z set to unity. The available energy during the scavenge period B is nearly equal to the minimum compressor work for this case. The blowdown work A is a bonus. For low boost pressure (p_{02}/p_{01}) the proportion of the energy available from the blowdown period is very much greater than for high boost pressures. It will be seen from expression (10.32) that the energy available from the blowdown period is independent of the scavenge ratio λ but is strongly dependent on the boost pressure ratio (p_{02}/p_{01}) as shown in Fig. 10.16. Since for the ideal case the minimum compressor work is approximately equal to the scavenge period available energy B, the excess energy A is available to be returned to the engine power output. This could be achieved by coupling the turbine to the engine. This is called compounding.

As losses appear in the system so the maximum available excess energy is reduced. We shall discuss this later.

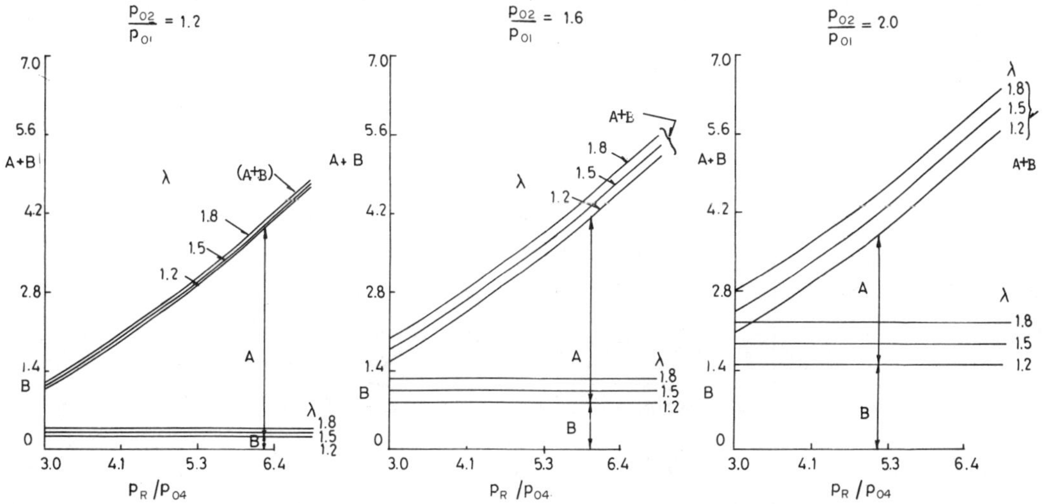

FIG. 10.16. Maximum available energy at turbine—two-stroke engine. B, energy from scavenge period; A, energy from blowdown period.

10.6.2 Four-stroke Engine

The pressure/crankangle diagram is shown in Fig. 10.14(b) and the pressure/volume diagram in Fig. 10.17. We shall examine in three parts: the exhaust blowdown, the exhaust stroke and valve overlap.

Exhaust Blowdown (Period R-S)

This is the same as the two-stroke model.

Thus

$$\left(\frac{W_T}{p_{04}V_R}\right)_{RS} = \frac{\left(\frac{p_R}{p_{04}}\right)}{\kappa-1}\left\{(1-xy)-\left(\frac{\kappa-1}{2}\right)(y-1)(1+x)+\kappa\left(\frac{R}{p_{04}}\right)^{\frac{1-\kappa}{\kappa}}\left(yx^{\frac{1}{\kappa}}-1\right)\right\}, \tag{10.32}$$

where $x = \frac{p_S}{p_R}$ and $y = \frac{V_S}{V_R}$.

Exhaust Stroke (Period S-C')

The exhaust stroke is assumed to be at constant pressure.

Once again we apply the control volume (a) to the engine cylinders and turbine, and (b) to the engine cylinders alone. The first law expressions then become:

Engine and turbine:

$$-dW_T - p\ dV = \frac{1}{\kappa-1}\ d(pV) + dm_4 h_{04}. \tag{10.47}$$

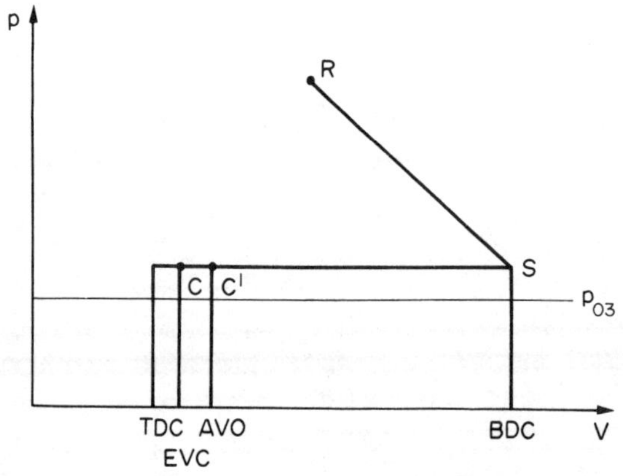

FIG. 10.17. Pressure/volume diagram.

Engine:

$$- p\ dV = \frac{1}{\kappa-1} d(pV) + dm_3 h_{03}. \qquad (10.48)$$

notice $dm_3 = dm_4$ = exhaust flow from engine,

$$\frac{h_{04}}{h_{03}} = \left(\frac{p_{04}}{p_{03}}\right)^{\frac{\kappa-1}{\kappa}} = \text{constant}.$$

This corresponds to isentropic expansion across the turbine.

From (10.48), since dp = 0,

$$dm_3 h_{03} = - \frac{\kappa}{\kappa-1} p\ dV.$$

From (10.47) the turbine work is

$$- dW_T = \frac{\kappa}{\kappa-1} p\ dV + dm_4 h_{04}. \qquad (10.49)$$

Now $$dm_4 h_{04} = dm_3 h_{04} = dm_3 h_{03} \left(\frac{h_{04}}{h_{03}}\right) = dm_3 h_{03} \left(\frac{p_{04}}{p_{03}}\right)^{\frac{\kappa-1}{\kappa}}.$$

Hence $$dm_4 h_{04} = -\left(\frac{p_{04}}{p_{03}}\right)^{\frac{\kappa-1}{\kappa}} \frac{\kappa}{\kappa-1} p\ dV.$$

Then (10.49) becomes

$$- dW_T = \frac{\kappa}{\kappa-1} p\ dV - \frac{\kappa}{\kappa-1} p\ dV \left(\frac{p_{04}}{p_{03}}\right)^{\frac{\kappa-1}{\kappa}}$$

or $$- dW_T = \frac{\kappa}{\kappa-1} \left(1-\left(\frac{p_{04}}{p_{03}}\right)^{\frac{\kappa-1}{\kappa}}\right) p\ dV.$$

Integrating between V_S and $V_{C'}$, we have

$$- W_T = \frac{\kappa}{\kappa-1} \left(1-\left(\frac{p_{04}}{p_{03}}\right)^{\frac{\kappa-1}{\kappa}}\right) p_S(V_{C'}-V_S) \qquad (10.50)$$

or, dividing by V_R and p_{04},

$$\frac{W_T}{p_{04}V_R} = \frac{\kappa}{\kappa-1}\left[1-\left(\frac{p_{04}}{p_{03}}\right)^{\frac{\kappa-1}{\kappa}}\right]\left(\frac{p_S}{p_{04}}\right)\left(\frac{V_S}{V_R}\right)\left(1-\frac{V_{C'}}{V_S}\right). \tag{10.51}$$

Now letting

$$w = \frac{V_{C'}}{V_S} \quad \text{and} \quad \frac{p_S}{p_{04}} = \frac{p_S}{p_{02}}\frac{p_{02}}{p_{04}},$$

if $\quad p_{01} = p_{04}$, then $\dfrac{p_S}{p_{04}} = \dfrac{p_S}{p_{02}}\dfrac{p_{02}}{p_{01}} = \alpha r,$

$$\frac{V_S}{V_R} = y, \quad \alpha = \frac{p_S}{p_{02}}, \quad r = \frac{p_{02}}{p_{01}}.$$

Then (10.51) becomes

$$\left(\frac{W_T}{p_{04}V_R}\right)_{SC'} = \frac{\kappa}{\kappa-1}\left[1-\left(\frac{p_{04}}{p_{03}}\right)^{\frac{\kappa-1}{\kappa}}\right]\alpha r y(1-w). \tag{10.52}$$

Valve Overlap (Period CC')

This is the same as the scavenge period for a two-stroke engine with changes to suit the different volumes.

$$W_T = \left[1-\left(\frac{p_{04}}{p_{03}}\right)^{\frac{\kappa-1}{\kappa}}\right]\left[\frac{\kappa}{\kappa-1}\,p_S\left(V_{C'}-V_C\right)+m_a h_{02}\right]. \tag{10.53}$$

Now m_a is the mass of air supplied during the overlap period. This is not the total mass of air to the engine.

Let $\quad \lambda' = \dfrac{V_{02}}{V_C}.$

This is the ratio of the volume of air supplied during the overlap period to the volume of the cylinder at exhaust valve closure, and

$$m_a h_{02} = \frac{\kappa\lambda'}{\kappa-1}\,p_{02}V_C.$$

Then, as before (for two-stroke period SC),

$$\alpha = \frac{p_S}{p_{02}}, \quad z = \frac{V_C}{V_R}, \quad y = \frac{V_S}{V_R}, \quad t = \frac{V_{C'}}{V_R},$$

$$\frac{W_T}{p_{04}V_R} = \frac{p_{02}}{p_{04}}\left(1-\left(\frac{p_{04}}{p_{03}}\right)^{\frac{\kappa-1}{\kappa}}\right) \frac{\kappa}{\kappa-1} \frac{V_C}{V_R}\left(\lambda'-\alpha\left(1-\frac{V_{C'}}{V_C}\right)\right).$$

Hence

$$\left(\frac{W_T}{p_{04}V_R}\right)_{C'C} = \left(\frac{p_{02}}{p_{04}}\right)\left(1-\left(\frac{p_{04}}{p_{03}}\right)^{\frac{\kappa-1}{\kappa}}\right)\left(\frac{z\kappa}{\kappa-1}\right)\left(\lambda'-\alpha\left(1-\frac{t}{z}\right)\right). \quad (10.54)$$

Compressor

For the compressor, as before,

$$W_C = m_{02}h_{02}\left(\frac{h_{01}}{h_{02}}\right)\left(\left(\frac{p_{02'}}{p_{01}}\ \frac{p_{02}}{p_{01}}\right)^{\frac{\kappa-1}{\kappa}} - 1\right).$$

Now $$m_{02}h_{02} = \frac{\kappa}{\kappa-1}\, p_{02}V_{02}.$$

We define

$$\lambda = \frac{V_{02}}{V_S},$$

where S corresponds to volume at bottom dead centre.

Then $$m_{02}h_{02} = \frac{\kappa\lambda}{\kappa-1}\, p_{02}V_S.$$

Now $$y = \frac{V_S}{V_R}.$$

Then $$m_{02}h_{02} = \frac{\kappa\lambda y}{\kappa-1}\, p_{02}V_R$$

and $$W_C = \frac{\kappa\lambda}{\kappa-1}\, y\, p_{02}V_R\left(\frac{h_{01}}{h_{02}}\right)\left(\left(\frac{p_{02'}}{p_{02}}\ \frac{p_{02}}{p_{01}}\right)^{\frac{\kappa-1}{\kappa}} - 1\right),$$

$$\left(\frac{W_C}{p_{04}V_R}\right) = \frac{n\lambda yr\kappa}{\kappa-1}\left(\left(\frac{r}{v}\right)^{\frac{\kappa-1}{\kappa}} - 1\right). \quad (10.55)$$

Summarizing for the three periods the maximum available energies are:

$$A = \left(\frac{W_T}{p_{04}V_R}\right)_{RS} = \frac{\left(\frac{p_R}{p_{04}}\right)}{\kappa-1}\left[(1-xy)-\left(\frac{\kappa-1}{2}\right)(y-1)(1+x)+\kappa\left(\frac{p_R}{p_{04}}\right)^{\frac{1-\kappa}{\kappa}}(yx^{\frac{1}{\kappa}}-1)\right]. \tag{10.32}$$

$$B = \left(\frac{W_T}{p_{04}V_R}\right)_{SC'} = \frac{\kappa\alpha ry}{\kappa-1}\left[1-\left(\frac{p_{04}}{p_{03}}\right)^{\frac{\kappa-1}{\kappa}}\right](1-w). \tag{10.52}$$

$$C = \left(\frac{W_T}{p_{04}V_R}\right)_{C'C} = \frac{zr\kappa}{\kappa-1}\left[\left[1-\left(\frac{p_{04}}{p_{03}}\right)^{\frac{\kappa-1}{\kappa}}\right]\left[\lambda' - \alpha\left(1-\frac{t}{z}\right)\right]\right]. \tag{10.54}$$

And the <u>minimum</u> compressor work is

$$\left(\frac{W_C}{p_{04}V_R}\right) = \frac{n\lambda yr\kappa}{\kappa-1}\left[\left(\frac{r}{v}\right)^{\frac{\kappa-1}{\kappa}}-1\right], \tag{10.55}$$

where

$$x = \frac{p_S}{p_R}, \quad y = \frac{V_S}{V_R}, \quad w = \frac{V_{C'}}{V_S}, \quad = \frac{p_S}{p_{02}}, \quad r = \frac{p_{02}}{p_{01}},$$

$$z = \frac{V_C}{V_R}, \quad t = \frac{V_{C'}}{V_R}, \quad n = \frac{T_{01}}{T_{02}},$$

$$\lambda' = \frac{\text{air flow for overlap only}}{\text{cylinder volume at EVC}} = \frac{V_{02}}{V_C},$$

$$\lambda = \frac{\text{total air supply}}{\text{cylinder volume at BDC}} = \frac{V_{02}}{V_S}.$$

The total energy available at the turbine is

$$\left(\frac{W_T}{p_{04}V_R}\right)_{tot} = \left(\frac{W_T}{p_{04}V_R}\right)_{RS} + \left(\frac{W_T}{p_{04}V_R}\right)_{SC'} + \left(\frac{W_T}{p_{04}V_R}\right)_{C'C} \tag{10.56}$$

$$= A + B + C. \tag{10.57}$$

In Fig. 10.18 the maximum available energies A, B, C at the turbine are shown for a typical four-stroke engine with some valve overlap. Typical values for the volume ratios are:

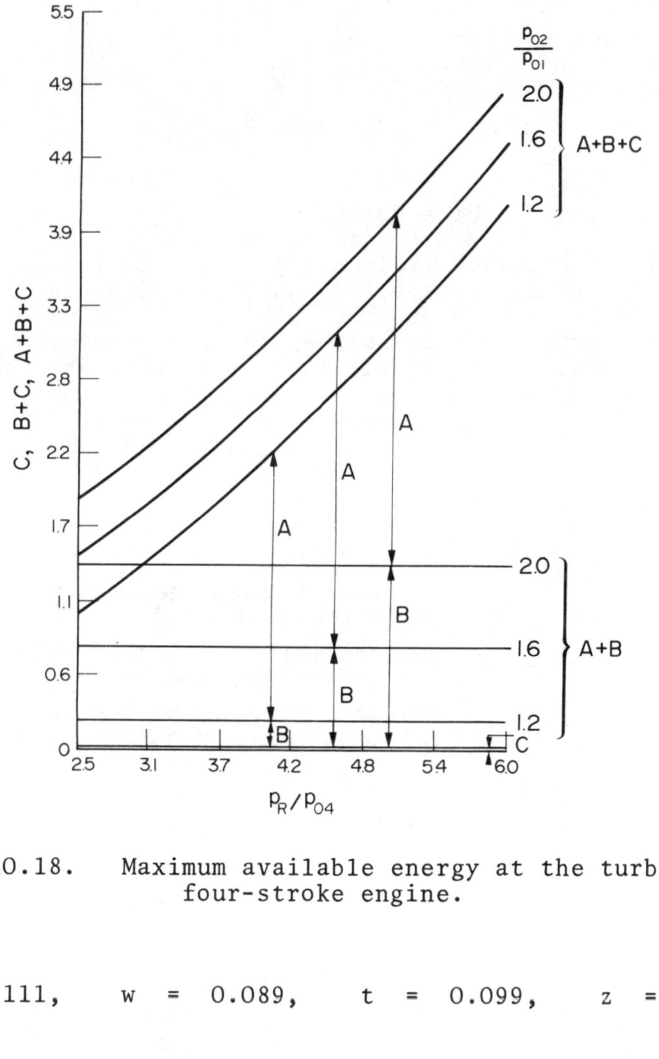

FIG. 10.18. Maximum available energy at the turbine—four-stroke engine.

$y = 1.111,\quad w = 0.089,\quad t = 0.099,\quad z = 0.085$

For the pressure ratio, $\alpha = 1.1$.

For the temperature ratio (assuming perfect after cooling), $n = 1.0$.

The scavenge ratios $\lambda' = 0.1(y/z)$, $\lambda = 1.1$.

The overlap energy C is negligible. Again the major component of the available energy is in the blowdown period A.

10.7 ACTUAL TURBOCHARGER SYSTEM

In the ideal systems discussed above

$$\frac{W_C}{p_{04}V_R} < \left(\frac{W_T}{p_{04}V_R}\right)_{tot}$$

and there is excess turbine work.

However, in practice it is not possible to locate the turbine, so that the valves or ports act as nozzles and a pipe is fitted between the cylinder head and the turbine. This causes throttling losses at the port or valves so that the energy available at the turbine is reduced. In Fig. 10.19 some results are shown for a

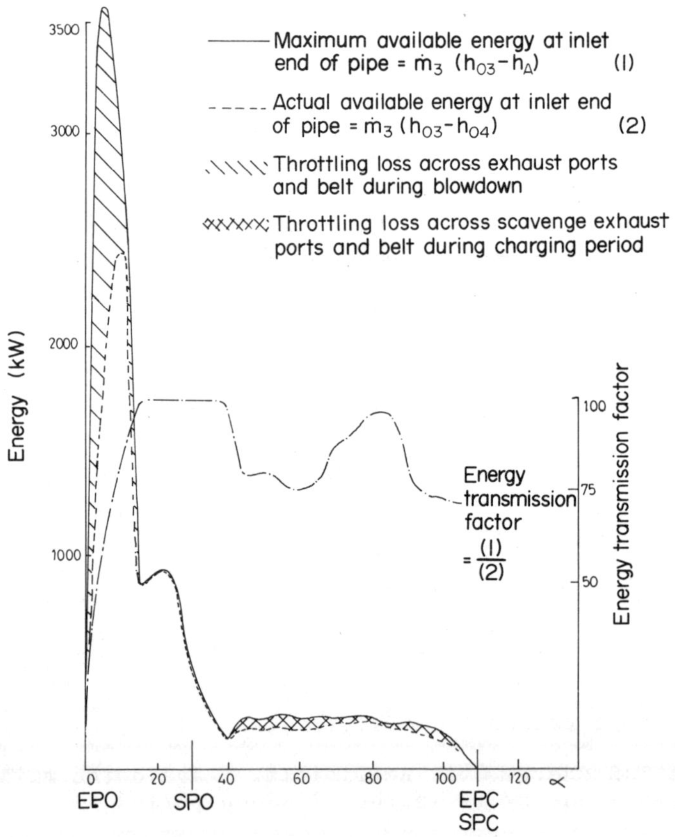

FIG. 10.19. Energy losses at entry to the exhaust system.

typical marine two-stroke engine exhaust port belt system. When the ports first open there is "wire-drawing" across the port. The losses gradually reduce as the pipe pressure approaches the cylinder pressure. When the scavenge and exhaust ports are both open there is a second period of loss. The energy transmission factor is defined as the ratio of the isentropic enthalpy drop in the pipe inlet to pressure p_{04} to the isentropic enthalpy drop corresponding to expansion from cylinder pressure $p_{C'}$.

The sudden release of high pressure gas into the pipe causes a pressure wave to be propagated along the pipe which is reflected at the turbine. Depending on the magnitude of the reflected wave, so the pressure levels will be different at the turbine end of the pipe from the cylinder end. The energy available for the turbine may or may not be less than the energy available at the inlet end of the pipe. To allow for the throttle losses at inlet and wave action effects in the pipe we define an exhaust pipe efficiency η_A given by

$$\eta_A = \frac{\text{available energy at the inlet to the turbine}}{\text{available energy for ideal turbocharger system}}$$

$$\eta_A = \frac{\left(W_T\right)_I}{\left(W_T\right)_{tot}} . \qquad (10.58)$$

We shall discuss the influence of the exhaust system design on the exhaust pipe efficiency later.

If the average turbine efficiency is η_T, then the work done by the turbine $\left(W_T\right)_{act}$ will be

$$\left(W_T\right)_{act} = \eta_T\left(W_T\right)_I \quad \text{or} \quad \left(W_T\right)_{act} = \eta_T\eta_A\left(W_T\right)_{tot} .$$

The actual compressor work will be W_C/η_C, where the compressor efficiency is η_C. If the mechanical efficiency of the turbocharger is η_M, then the energy balance for the turbocharger is

$$\eta_M\eta_T\eta_A\left(W_T\right)_{tot} = \frac{W_C}{\eta_C} \quad \text{or} \quad \eta_M\eta_T\eta_C\eta_A = \frac{W_C}{\left(W_T\right)_{tot}} .$$

This expression can be expressed in the non-dimensional terms of the ideal system as

$$\eta_M\eta_T\eta_C\eta_A = \frac{\left(\dfrac{W_C}{p_{04}V_R}\right)}{\left(\dfrac{W_T}{p_{04}V_R}\right)_{tot}} .$$

If we set K equal to the product $\eta_M\eta_T\eta_C\eta_A$,

$$K = \frac{\left(\dfrac{W_C}{p_{04}V_R}\right)}{\left(\dfrac{W_T}{p_{04}V_R}\right)_{tot}} \quad . \tag{10.59}$$

Thus the ratio of the ideal compressor work to the maximum available energy at the inlet end of the exhaust pipe is equal to the product of the turbocharger efficiency η_{TC} and the exhaust pipe efficiency η_A. The parameter K is called the matching coefficient.

$$K = \eta_{TC}\eta_A = \eta_M\eta_T\eta_C\eta_A . \tag{10.60}$$

Using the results given in Figs. 10.14 and 10.16 we can plot the matching coefficients for the two- and four-stroke engine examples. In Fig. 10.20 the two-stroke engine matching coefficient is shown. It will be observed that for a given release pressure (p_R/p_{04}) higher coefficients are required for higher boosts (p_{02}/p_{01}) and scavenge ratio λ. The same trends are also observed for four-stroke engines (Fig. 10.21). The release pressure is related to the engine

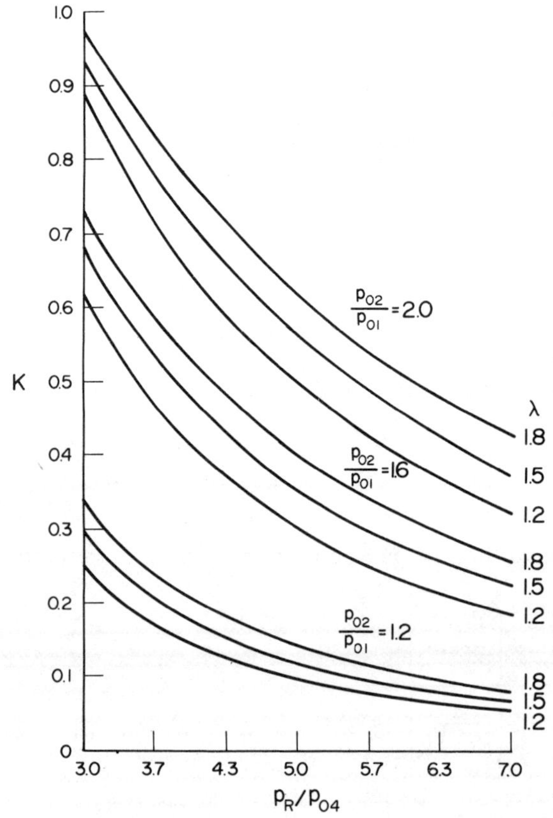

FIG. 10.20. Matching coefficient K for two-stroke engine.

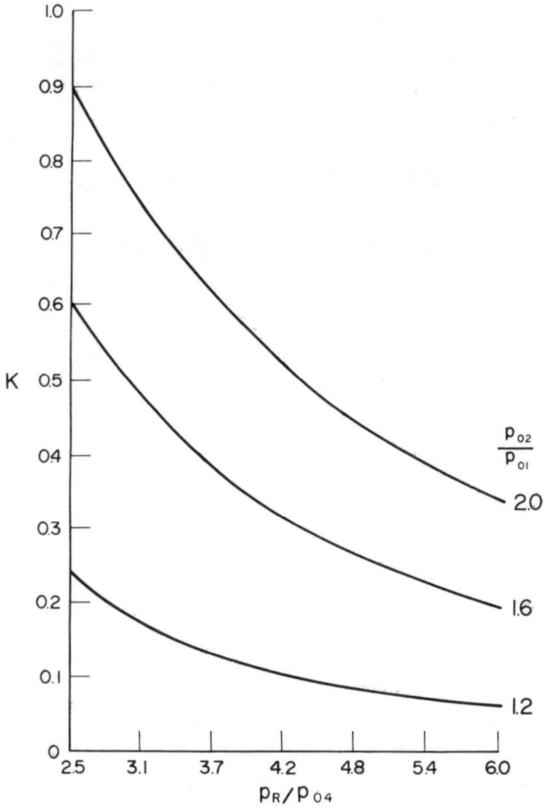

FIG. 10.21. Matching coefficient K for four-stroke engine.

power output. The latter is related to the thermal efficiency, air/fuel ratio and boost. It is possible to transform the graphs shown in Figs. 10.20 and 10.21 to matching charts (Figs. 10.22 and 10.23) (see Benson and Horlock(6) and Smalley(7)). It will be seen from the matching charts that there is one only matching coefficient K for given boost (p_{02}/p_{01}), exhaust temperature ratio (T_{03}/T_{01}) and power (b.m.e.p.). If one or more component efficiency in the product $\eta_A\eta_T\eta_C$ is increased, one can obtain the same power (b.m.e.p.) with lower exhaust temperature (T_{03}/T_{01}) and higher boost (p_{02}/p_{01}).

One of the important components of the turbocharger system is the exhaust system. We shall now examine the factors which influence exhaust system efficiency.

10.8 EFFICIENCY OF EXHAUST SYSTEMS

In the previous discussion we have calculated the maximum available energy to drive the turbocharger turbine. In this case there will be no losses in the exhaust system and in the optimum

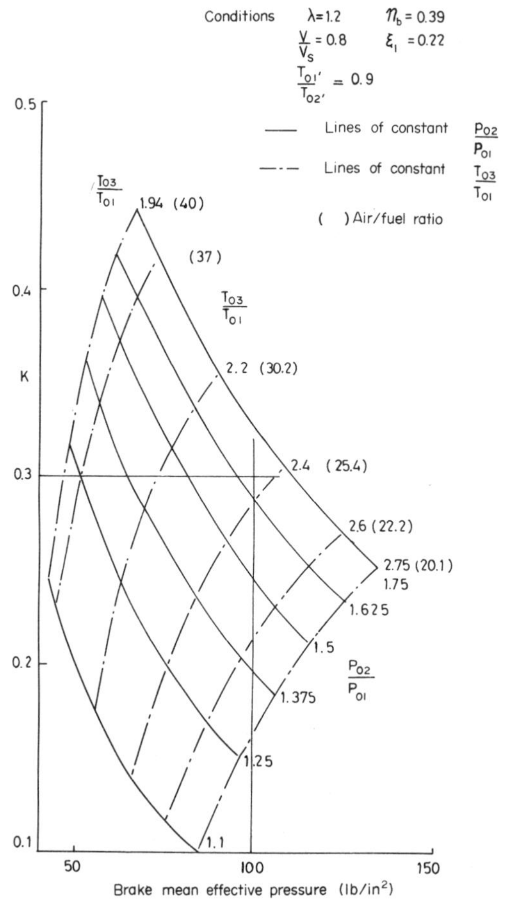

FIG. 10.22. Matching chart for scavenge ratio 1.2. η_A, exhaust availability factor; η_C compressor efficiency; η_T turbine efficiency. (From Benson and Horlock[4] by courtesy of CIMAC).

case $p_{03} = p_{cyl}$. That is, the stagnation pressure in the exhaust pipe equals the stagnation pressure in the cylinder. Thus a plot of stagnation pressure in the exhaust pipe will follow the cylinder pressure and we shall have the ideal diagram shown in Fig. 10.24(a).

For small compact exhaust systems we will have a pulse following the ideal shape (Fig. 10.24(b)). For a large receiver exhaust system we shall have almost constant pressure ahead of the turbine. The latter type of system is called a constant pressure charging system and the former a pulse charging system. We shall first discuss the efficiency of constant pressure charging systems, then pulse charging systems.

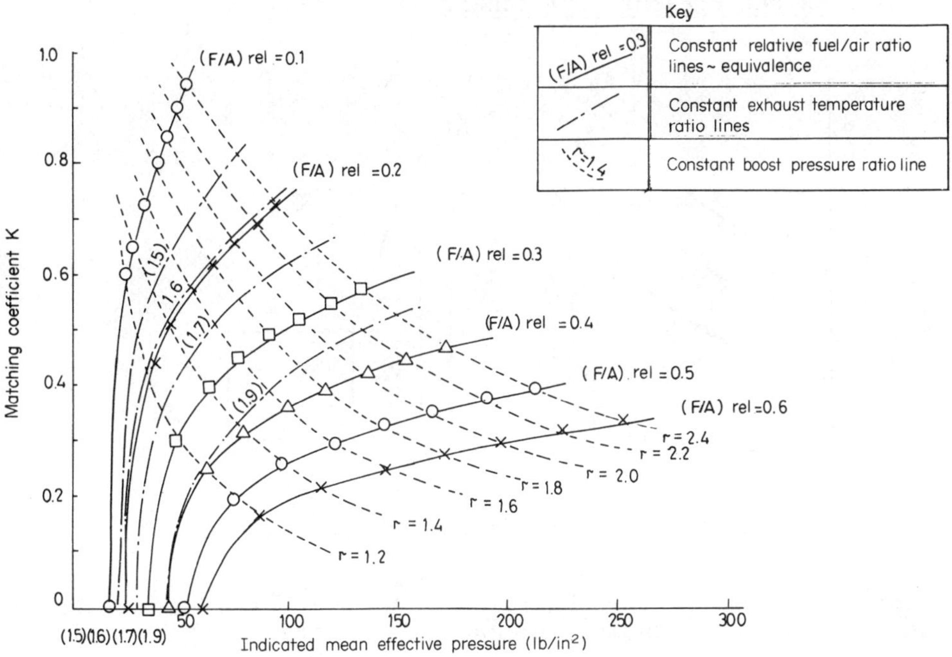

FIG. 10.23. Matching chart for engine speed = 1800 rev/min, four-stroke engine.

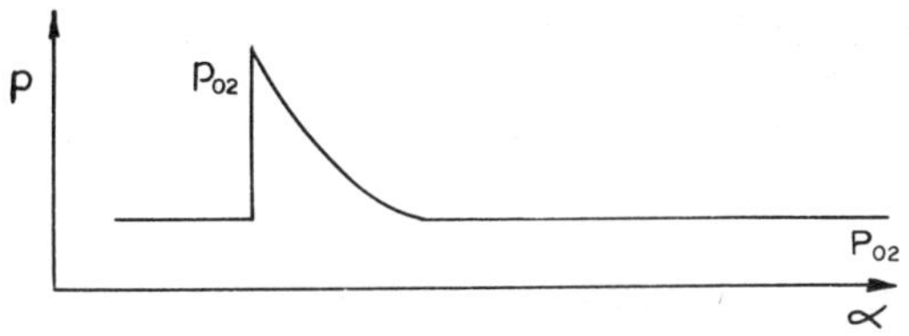

(a) Ideal pipe diagram ahead of turbine

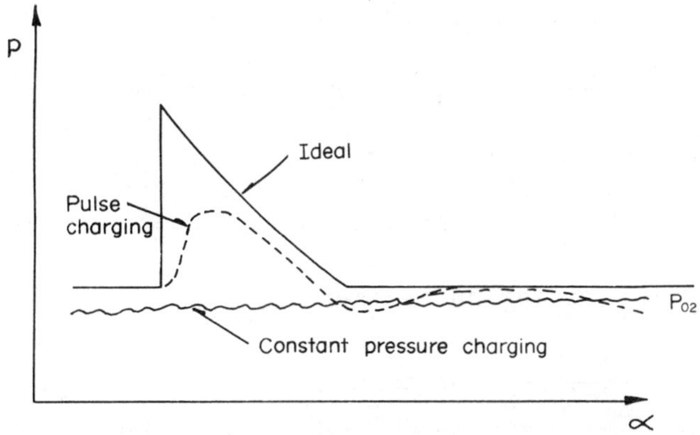

(b) Actual pipe diagram ahead of turbine

FIG. 10.24. Exhaust pipe: pressure diagrams ahead of turbine.

10.8.1 Constant Pressure Turbocharging

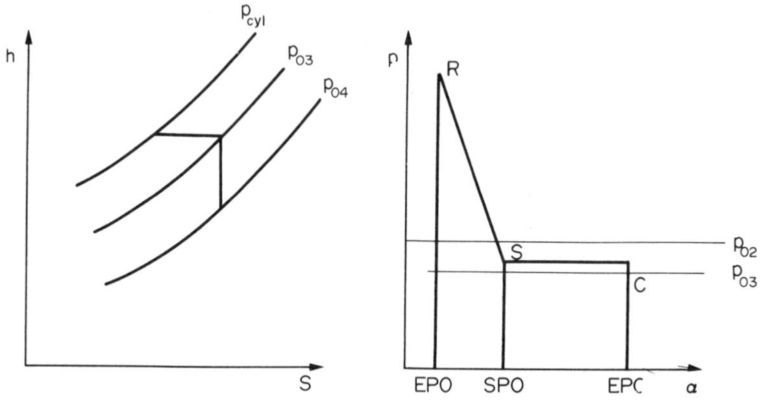

FIG. 10.25. Constant pressure charging.

Constant pressure turbocharging is usually used on two-stroke engines and we shall confine our analysis to this case.

We assume that the gases are throttled across the exhaust ports from p_{cyl} to p_{03} (Fig. 10.25). Once again we examine the two separate periods.

Exhaust Blowdown (Period R-S)

Apply the first law to a control volume including the cylinder and turbine then

$$- dW_T - p\, dV = \frac{1}{\kappa-1} d(pV) + dm_4 h_{04}. \tag{10.61}$$

Now $$dm_4 = - dm_{cyl} = - dm.$$

Hence $$- dW_T = \frac{1}{\kappa-1} d(pV) + p\, dV - dm h_{04}. \tag{10.62}$$

Now $$h_{04} = h_{cyl}\left(\frac{p_{04}}{p_{03}}\right)^{\frac{\kappa-1}{\kappa}} = h\left(\frac{p_{04}}{p_{03}}\right)^{\frac{\kappa-1}{\kappa}}.$$

In the cylinder the expansion is isentropic.

Hence $$\frac{T}{\rho^{\kappa-1}} = \text{constant}$$

or $$h = \text{constant} \times \rho^{\kappa-1}.$$

Hence $$\frac{dh}{h} = (\kappa-1)\frac{d\rho}{\rho} \quad \text{or} \quad \frac{d\rho}{\rho} = \frac{1}{\kappa-1}\frac{dh}{h}.$$

In the cylinder:

$$m = \rho V,$$

$$dm = \rho\, dV + V\, d\rho = \rho\, dV + \frac{\rho V}{\kappa-1}\frac{dh}{h},$$

$$dm = \rho\, dV + \frac{m}{(\kappa-1)}\frac{dh}{h},$$

$$(\kappa-1)h\, dm = (\kappa-1)h\, \rho\, dV + m\, dh,$$

$$(\kappa-1)h\, dm = (\kappa-1)\, hm\frac{dV}{V} + m\, dh. \tag{10.63}$$

Now $$\frac{p}{\rho} = RT = \left(C_p - C_v\right)T = C_p T\left(1 - \frac{1}{\kappa}\right) = \left(\frac{\kappa-1}{\kappa}\right)h.$$

Hence $$\kappa p = \rho h(\kappa-1)$$

or $$\kappa pV = \rho V h(\kappa-1) = mh(\kappa-1).$$

Substitute for $mh(\kappa-1)$ in (10.63),

$$(\kappa-1)h\, dm = \kappa p\, dV + m\, dh,$$

therefore

$$h\, dm + m\, dh = \kappa h\, dm - \kappa p\, dV$$

or $$\frac{1}{\kappa}\, d(mh) = h\, dm - p\, dV.$$

Now $$mh = \left(\frac{\kappa}{\kappa-1}\right)pV.$$

Hence $$\frac{1}{\kappa-1}\, d(pV) = h\, dm - p\, dV$$

or $$h\, dm = \frac{1}{\kappa-1}\, d(pV) + p\, dV. \tag{10.64}$$

Now substitute for h_{04} into (10.62),

$$-\, dW_T = \frac{1}{\kappa-1}\, d(pV) + p\, dV - \left(\frac{p_{04}}{p_{03}}\right)^{\frac{\kappa-1}{\kappa}} h\, dm,$$

and substitute from (10.64) for h dm,

$$- dW_T = \frac{1}{\kappa-1}\, d(pV) + p\, dV - \left(\frac{p_{04}}{p_{03}}\right)^{\frac{\kappa-1}{\kappa}} \left(\frac{1}{\kappa-1}\, d(pV) + p\, dV\right)$$

or

$$- dW_T = \left(1-\left(\frac{p_{04}}{p_{03}}\right)^{\frac{\kappa-1}{\kappa}}\right)\left(\frac{1}{\kappa-1}\, d(pV) + p\, dV\right).$$

Integrate between R and S and observe that for linear variation of p with V,

$$\int_R^S p\, dV = \left(\frac{p_R+p_S}{2}\right)\left(V_S-V_R\right).$$

Then

$$- W_T = \left(1-\left(\frac{p_{04}}{p_{03}}\right)^{\frac{\kappa-1}{\kappa}}\right)\left(\frac{p_SV_S-p_RV_R}{\kappa-1} + \left(\frac{p_R+p_S}{2}\right)\left(V_S-V_R\right)\right)$$

$$- W_T = p_RV_R\left(1-\left(\frac{p_{04}}{p_{03}}\right)^{\frac{\kappa-1}{\kappa}}\right)\left\{\left(\frac{p_SV_S}{p_RV_R} -1\right)\left(\frac{1}{\kappa-1}\right) + \tfrac{1}{2}\left(1+ \frac{p_S}{p_R}\right)\left(\frac{V_S}{V_R} -1\right)\right\}. \qquad (10.65)$$

Divide by $p_{04}V_R$ and change sign,

$$\frac{W_T}{p_{04}V_R} = \frac{1}{(\kappa-1)}\left(\frac{p_R}{p_{04}}\right)\left(1-\left(\frac{p_{04}}{p_{03}}\right)^{\frac{\kappa-1}{\kappa}}\right)\left(1-\left(\frac{p_S}{p_R}\right)\left(\frac{V_S}{V_R}\right) + \frac{\kappa-1}{2}\left(1+ \frac{p_S}{p_R}\right)\left(1- \frac{V_S}{V_R}\right)\right). \qquad (10.66)$$

Now

$$x = \frac{p_S}{p_R}, \qquad y = \frac{V_S}{V_R}.$$

Then

$$\left(\frac{W_T}{p_{04}V_R}\right)_{RS} = \frac{1}{(\kappa-1)}\left(\frac{p_R}{p_{04}}\right)\left(1-\left(\frac{p_{04}}{p_{03}}\right)^{\frac{\kappa-1}{\kappa}}\right)\left(1-xy-\left(\frac{\kappa-1}{2}\right)(1+x)(y-1)\right). \qquad (10.67)$$

For the <u>scavenge period S-C</u> we use the same expression as for the scavenge period for the ideal system (10.40):

$$\left(\frac{W_T}{p_{04}V_R}\right)_{SC} = \left(\frac{p_{02}}{p_{04}}\right)\left(1-\left(\frac{p_{04}}{p_{03}}\right)^{\frac{\kappa-1}{\kappa}}\right)\left(\frac{z\kappa}{\kappa-1}\right)\left(\lambda-\alpha\left(1- \frac{y}{z}\right)\right). \qquad (10.40)$$

The total available turbine work is now

$$\frac{W_T}{p_{04}V_R} = \left(\frac{W_T}{p_{04}V_R}\right)_{RS} + \left(\frac{W_T}{p_{04}V_R}\right)_{SC} . \tag{10.68}$$

The maximum available energy will be less than the ideal and the maximum efficiency of the exhaust pipe for constant pressure charging is therefore given by

$$\eta_A = \frac{\left(\frac{W_T}{p_{04}V_R}\right)_{RS_{cp}} + \left(\frac{W_T}{p_{04}V_R}\right)_{SC}}{\left(\frac{W_T}{p_{04}V_R}\right)_{RS_{ideal}} + \left(\frac{W_T}{p_{04}V_R}\right)_{SC}} , \tag{10.69}$$

where $\left(\frac{W_T}{p_{04}V_R}\right)_{RS_{cp}}$ corresponds to the constant pressure exhaust blowdown, (10.67)

$\left(\frac{W_T}{p_{04}V_R}\right)_{RS_{ideal}}$ corresponds to the ideal exhaust blowdown. (10.32)

In Fig. 10.26 the blowdown energy A and the scavenge period energy B are shown. There is a noticeable reduction in available energy A due to the throttling losses during blowdown. The maximum possible efficiency of the exhaust system is shown in Fig. 10.27. Hence it will be seen that the maximum possible efficiency increases with boost (p_{02}/p_{01}). The results shown in Fig. 10.27 can be

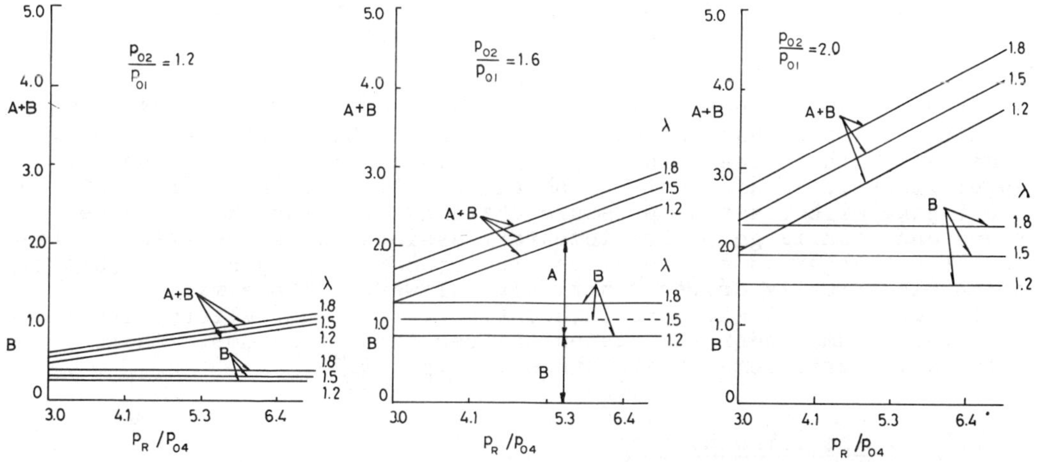

FIG. 10.26. Maximum available energy at the turbine constant pressure charging.

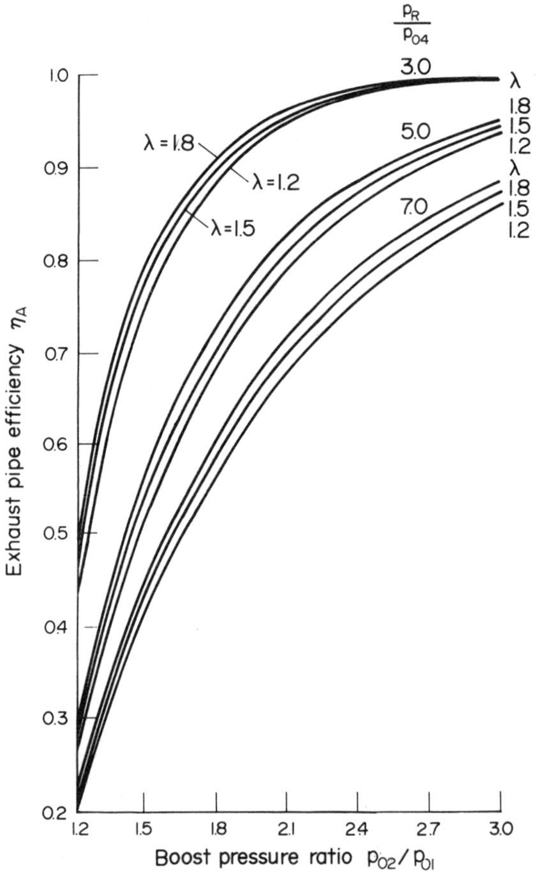

FIG. 10.27. Maximum possible exhaust pipe efficiency for constant pressure systems.

related to the trapped air/fuel ratio as shown in Fig. 10.28. These results indicate that the maximum possible exhaust pipe efficiency decreases with decrease in air/fuel ratio and decreases in boost. In practice it is difficult to obtain a self-sustained air supply to the engine with constant pressure charging over the whole engine speed and load range. Supplementary assistance is required. This can be either by an engine-driven reciprocating pump or Roots blower or an electrically driven centrifugal blower. These may be in series or parallel with the turbocharger. Another alternative is to seal the volume under the piston and use this as a pump. Some alternative arrangements are shown in Fig. 10.29.

10.8.2 Pulse Turbocharging

When the exhaust valve or ports are open, the high pressure gas causes a pressure wave to be propagated into the exhaust system. We

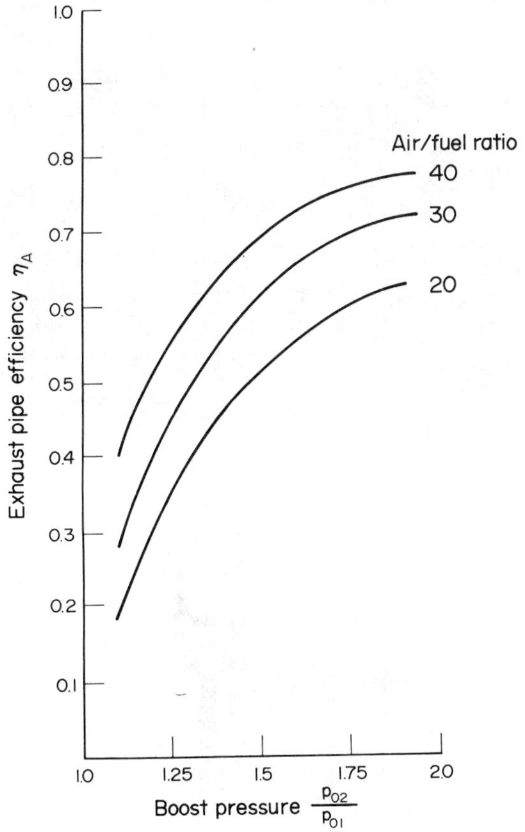

FIG. 10.28. Maximum possible exhaust pipe efficiency for constant pressure supercharge systems.

can study this phenomenon by applying the general equations of motion (2.71) and (2.78). A full analysis is extremely complex; we shall confine our remarks to a simplified outline.

A mathematical solution to (2.71) and (2.78) may be obtained through the theory of characteristics. This solution applies along lines called characteristics.[6,14] There are two sets of characteristics; namely, position characteristics and state characteristics.[7,14] We can devise a chart which is a pictorial presentation of the pressure waves in a position diagram (Fig. 10.30). This has an x-t, a length/time, co-ordinate system. The slope of lines in this diagram will be given by

$$\frac{dx}{dt} = u \pm a. \tag{10.70}$$

Along these lines the particle velocities u and the local speed of sound a will vary according to the expression

$$\frac{da}{du} = \mp \frac{\kappa-1}{2}. \tag{10.71}$$

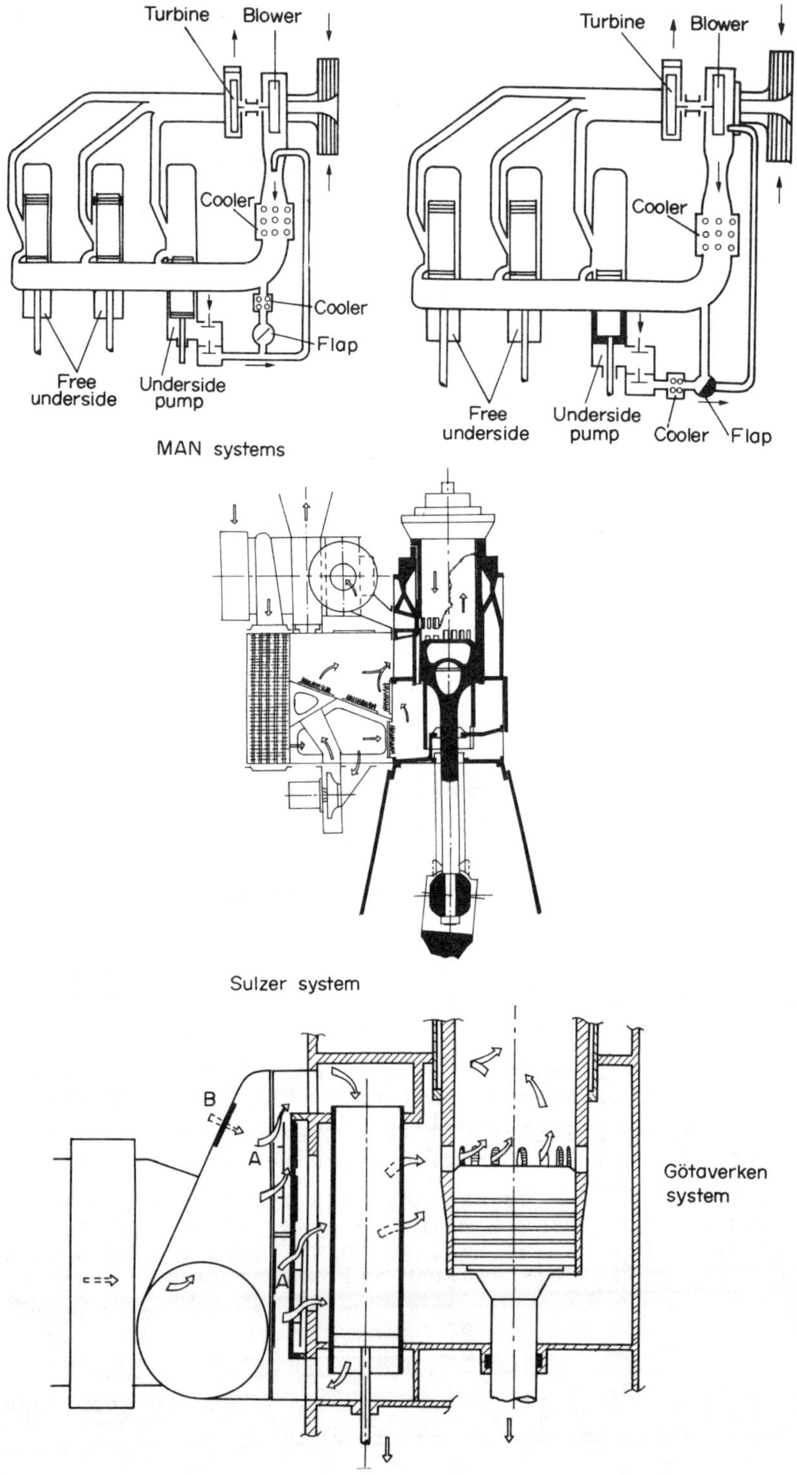

FIG. 10.29. Turbocharging systems for two-stroke engines.

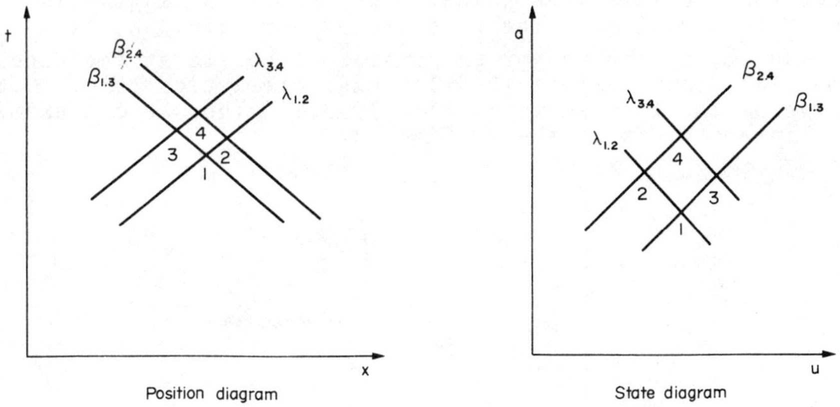

FIG. 10.30. Characteristic diagrams.

In Fig. 10.30 are shown the characteristic diagrams. Along the line $\beta_{1.3}$ in the position diagram the state of the gas changes from 1 to 3 in the state diagram. Thus at the point x_1, t_1 the state is a_1, u_1 and at the point x_2, t_2 the state is a_2, u_2.

The λ, β lines in the position diagrams are the wave lines and represent pressure waves being propagated along a duct to the right and left, respectively.

Equations (10.70) and (10.71) are solutions of the basic equations (2.71) and (2.78). We normally use the non-dimensional groups

$$A = \frac{a}{a_{ref}} = \left(\frac{p}{p_{ref}}\right)^{\frac{\kappa-1}{2\kappa}}, \quad U = \frac{u}{a_{ref}}, \quad Z = \frac{a_{ref}t}{L_{ref}}, \quad X = \frac{x}{L_{ref}}.$$

The reference conditions are shown in the entropy diagram (Fig. 10.31); L_{ref} is a pipe length.

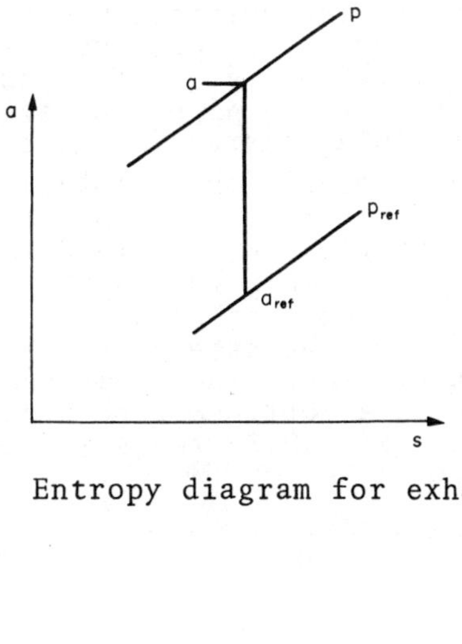

FIG. 10.31. Entropy diagram for exhaust pipe.

The state diagram is now an A-U diagram and since $A = (p/p_{ref})^{(\kappa-1)/2\kappa}$ A defines the pressure at a point. The position diagram is a Z-X diagram. This will give the pressure at any point and time. A turbine may be represented by an equivalent nozzle at the pipe end.[7] It is possible to transform the flow characteristics (Fig. 10.6(a)) to a curve of the form shown in Fig. 10.32. Thus we can examine the wave interactions at the turbine.

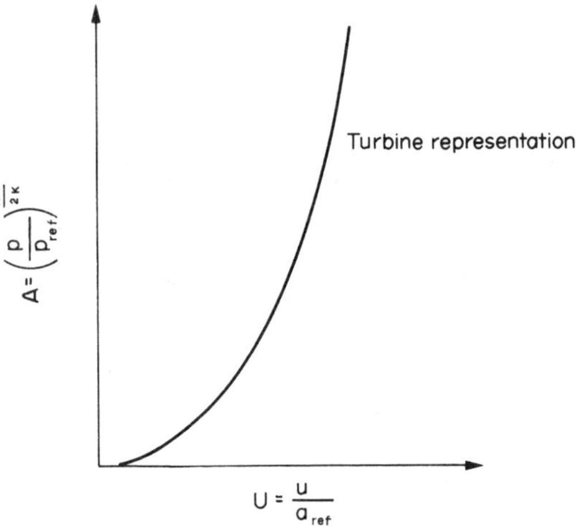

FIG. 10.32. Representation of a turbine on a state diagram.

In Fig. 10.33 a simple calculation of the wave interaction at the turbine is shown. The pressure wave at the inlet end of the pipe is represented by the points 1, 2, 4, 7, 11; this is reflected at the turbine where the pressure will be represented by the points 1, 3, 6, 10, 15. Notice in this calculation that there is a pressure rise at the turbine end of the pipe. Whether the pressure rises or falls will depend on the area ratio of the turbine nozzles to pipe area, the pipe length and engine speed. The magnitude of the inlet pressure will depend on the ratio of the port area (or valve) area to pipe area. The smaller the pipe length and the smaller the pipe diameter, the greater in general will be the pipe pressure pulses. But there is an optimum below which the effects produce a reduction in performance. If we calculate the available energy at the turbine end of the pipe and divide it by the ideal available energy at the inlet energy, we have the exhaust pipe efficiency. Benson and Woods[8] have suggested an alternative method of assessing the optimum proportion of an exhaust system. The maximum available energy at any instant is $\dot{m}\Delta h_{is}$, where $\dot{m}$ is the flow through the turbine and Δh_{is} the isentropic enthalpy drop. By suitably non-dimensionalizing this parameter the effect of pipe diameter and nozzle area may be assessed. In Fig. 10.34 the effect of pipe size on available energy is shown. It is quite clear that there is an optimum pipe area to cylinder cross-sectional area. In Fig. 10.35 the so-called nozzle area

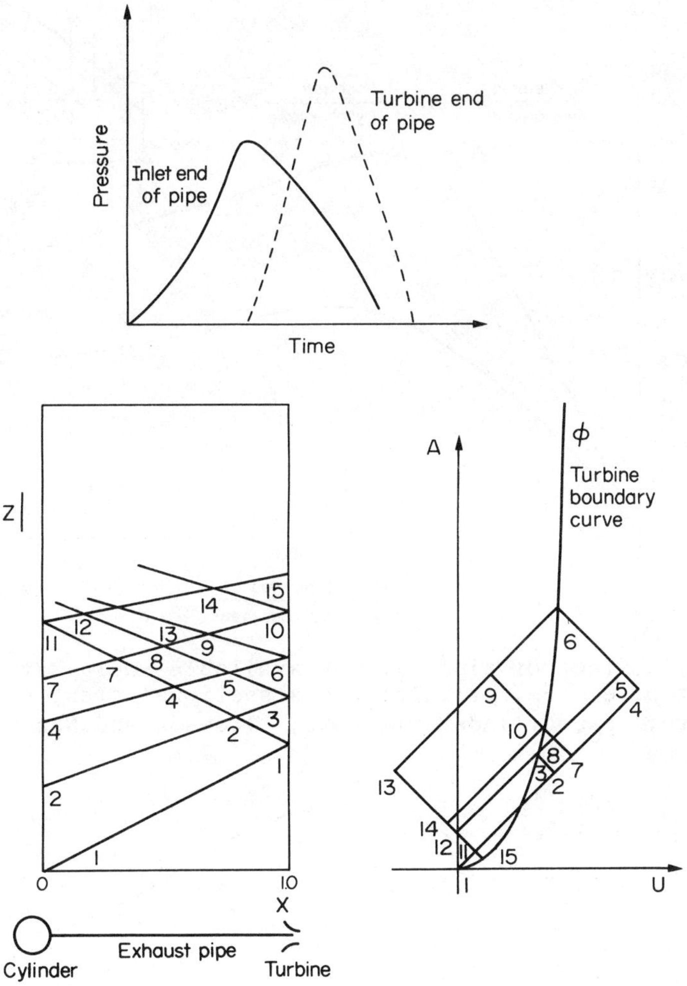

FIG. 10.33. Wave action at a turbine.

ratio is shown. This is the ratio of the turbine nozzle area to pipe cross-sectional area. Once again there is a clear optimum range.

Pulse turbocharging systems operate with varying enthalpy drop Δh_{is} across the turbine. We showed in Fig. 10.8 that this results in variable instantaneous turbine efficiencies η_{TS}.

The actual turbine work will be

$$\left(W_T\right)_{act} = \int \eta_{TS} \text{ in } (\Delta h_{is})dt.$$

The maximum turbine work will be

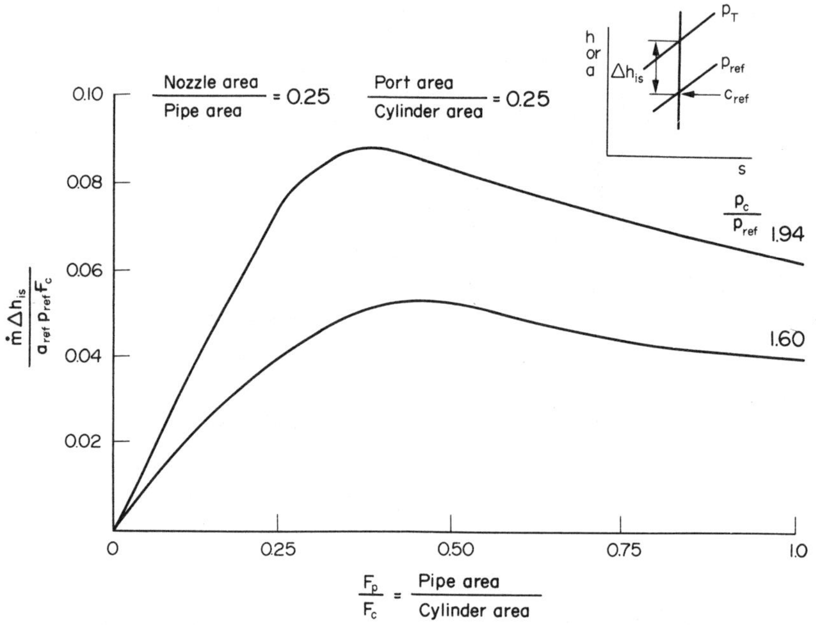

FIG. 10.34. Effect of pipe size on available energy at turbine end of exhaust pipe. F_C, cylinder cross-sectional area, p_T, turbine pressure; p_C, cylinder pressure; (Benson and Woods.[8])

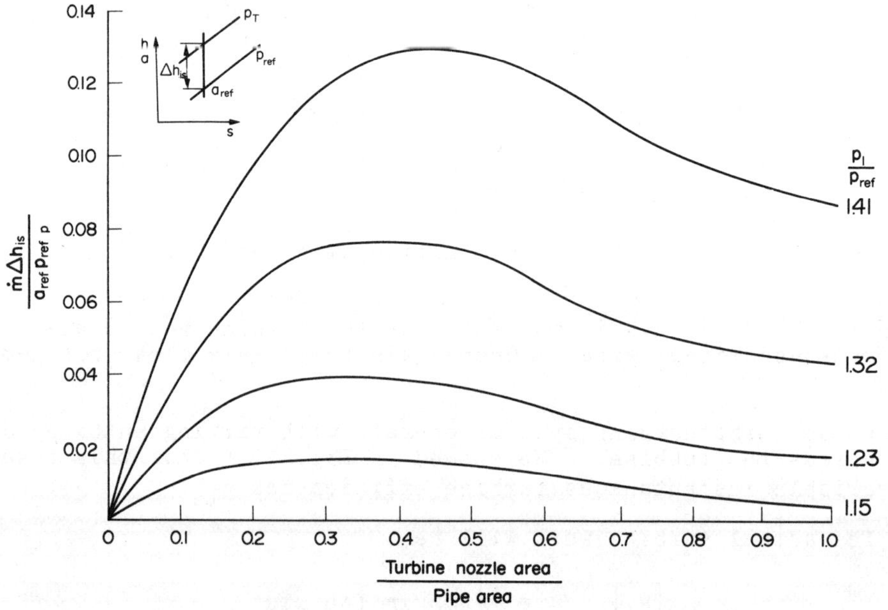

FIG. 10.35. Effect of nozzle area/pipe area ratio on available energy at nozzle end of exhaust pipe. F_p, pipe cross-sectional area; p_I, pressure at inlet end of exhaust pipe; p_T pressure at turbine end of exhaust pipe; (Benson and Woods.[8]).

$$\left(W_T\right)_{max} = \dot{m}\left(\Delta h_{is}\right)dt.$$

Then if we define the average turbine efficiency as

$$\eta_T = \frac{\int \eta_{TS}\ \dot{m}\Delta h_{is}dt}{\int \dot{m}\Delta h_{is}dt}$$

the actual turbine work will be

$$\left(W_T\right)_{act} = \eta_T\left(W_T\right)_{max}.$$

Now $\left(W_T\right)_{max}$ corresponds to the maximum available energy at the turbine inlet $\left(W_T\right)_I$ in expression (10.58). Then

$$\left(W_T\right)_{act} = \eta_T\left(W_T\right)_I$$

and $$\left(W_T\right)_I = \eta_A\left(W_T\right)_{tot} \quad \text{from (10.58).}$$

Hence $$\left(W_T\right)_{act} = \eta_T\eta_A\left(W_T\right)_{tot}$$

or $$\frac{\left(W_T\right)_{act}}{\left(W_T\right)_{tot}} = \eta_T\eta_A. \tag{10.72}$$

The product $\eta_T\eta_A$ is a measure of the efficiency of the exhaust/turbine system.

A constant pressure system can be designed to operate at the optimum turbine efficiency $(\eta_T)_{CP}$. This will be generally greater than the average efficiency $(\eta_T)_P$ for a pulse system, so that for an effective pulse-charging system the exhaust pipe efficiency $(\eta_A)_P$ must be greater for the pulse system than for the constant pressure system $(\eta_A)_{CP}$.

In practice the higher the boost the higher the exhaust efficiency $(\eta_A)_{CP}$; this will be clear from Fig. 10.27. Modern two-stroke diesels operate with constant pressure systems at full load and speed. However, as the speed falls, and hence the boost, the pipe efficiency η_A falls and supplementary systems are required to provide air (Fig. 10.30). On the other hand, well-designed pulse-charging systems can operate over the whole speed and load range without assistance. The major problem with a pulse-charging system is the necessity to avoid exhaust pipe interference between the cylinders. The methods outlined in Fig. 10.33 can be used to eliminate interference effects; these methods are described elsewhere by Benson.[9-11]

10.9 MATCHING TURBOCHARGER TO ENGINE

There are a number of methods for matching the turbocharger to the engine. We shall describe one method. Typical compressor characteristics are shown in Fig. 10.35a(a). It is convenient to replace the efficiency η_C data by a power output expression. The compressor power is

$$\dot{W}_C = \frac{\dot{m}_C C_p}{\eta_C}\left(T_{02s} - T_{01}\right).$$

This can be rearranged to give

$$\dot{W}_C = \dot{m}_C \frac{\kappa}{\kappa-1} R \frac{T_{01}}{\eta_C}\left(\left(\frac{p_{02}}{p_{01}}\right)^{\frac{\kappa-1}{\kappa}} - 1\right)$$

or

$$\frac{\dot{W}_C}{p_{01}\sqrt{T_{01}}} = \frac{1}{\eta_C}\left(\frac{\dot{m}_C\sqrt{T_{01}}}{p_{01}}\right)\left(\left(\frac{p_{02}}{p_{01}}\right)^{\frac{\kappa-1}{\kappa}} - 1\right)\left(\frac{\kappa R}{\kappa-1}\right). \quad (10.73)$$

Curves of $\left(\frac{\dot{W}_C}{p_{01}\sqrt{T_{01}}}\right)$ against $\left(\frac{p_{02}}{p_{01}}\right)$ are shown in Fig. 10.37(d).

For the turbine the power is

$$\dot{W}_T = \eta_T \dot{m}_T C_p\left(T_{03} - T_{04s}\right)$$

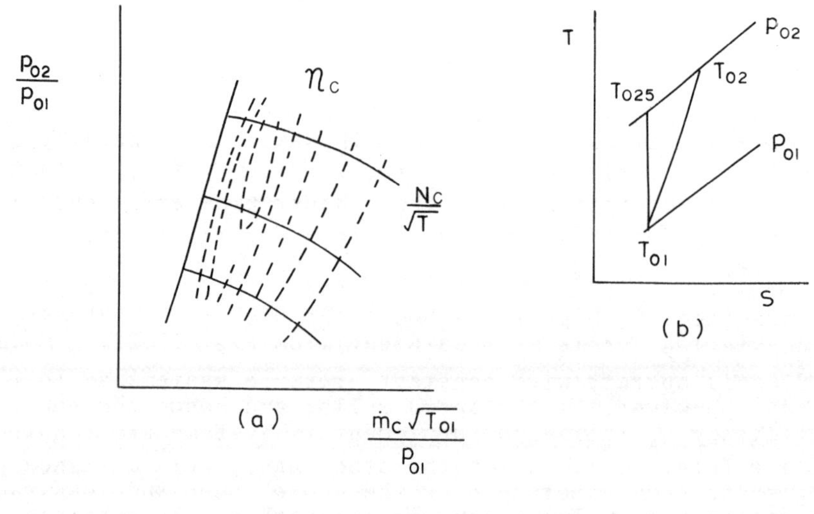

FIG. 10.35. Compressor characteristics.

or $$\dot{W}_T = \eta_T \dot{m} C_p T_{03}\left[1-\left(\frac{p_{04}}{p_{03}}\right)^{\frac{\kappa-1}{\kappa}}\right]$$

or $$\frac{\dot{W}_T}{\dot{m}\ T_{03}} = \eta_T C_p\left[1-\left(\frac{p_{04}}{p_{03}}\right)^{\frac{\kappa-1}{\kappa}}\right]. \qquad (10.74)$$

The turbine characteristics are normally given in the form shown in Fig. 10.36(a) and (c). The blade speed ratio u/c_{TS} can be rearranged as follows:

$$\frac{u}{c_{TS}} = \frac{\pi\ D\ N_T}{\left[2C_p\left(T_{03}-T_{04s}\right)\right]^{\frac{1}{2}}} = \frac{\pi\ D\ N_T}{2C_pT_{03}\left[1-\left(\frac{p_{04}}{p_{03}}\right)^{\frac{\kappa-1}{\kappa}}\right]^{\frac{1}{2}}}$$

$$\frac{u}{c_{TS}} = \frac{\pi\ D}{\left(2C_p\right)^{\frac{1}{2}}}\ \frac{\frac{N_T}{\sqrt{T_{03}}}}{\left[1-\left(\frac{p_{04}}{p_{03}}\right)^{\frac{\kappa-1}{\kappa}}\right]^{\frac{1}{2}}}, \qquad (10.75)$$

where D = rotor diameter.

The mass flow parameter for the turbine is expressed in the form

$$\frac{\dot{m}_T\sqrt{T_{03}}}{p_{04}} = \frac{\dot{m}_T\sqrt{T_{03}}}{p_{03}}\ \frac{p_{03}}{p_{04}} \qquad (10.76)$$

and the pressure/mass flow characteristics Fig. 10.36(a) are transformed to the form shown in Fig. 10.37 (e). The turbine work parameter (10.74) is evaluated from the pressure/mass flow curves Fig. 10.37(e), and the efficiency curve Fig. 10.36(c) through the expression (10.75) to give the curves shown in Fig. 10.37(f).

The matching procedure is a trial and error calculation. One can start with a given boost pressure ratio (p_{01}/p_{02}) and determine the air flow rate $\dot{m}_a$ from Fig. 10.37(a), or, given the flow rate $\dot{m}_a$, determine the boost. Either way the air/fuel ratio A/F is obtained from the fuel rate $\dot{m}_f$,

$$A/F = \frac{\dot{m}_a}{\dot{m}_f}. \qquad (10.77)$$

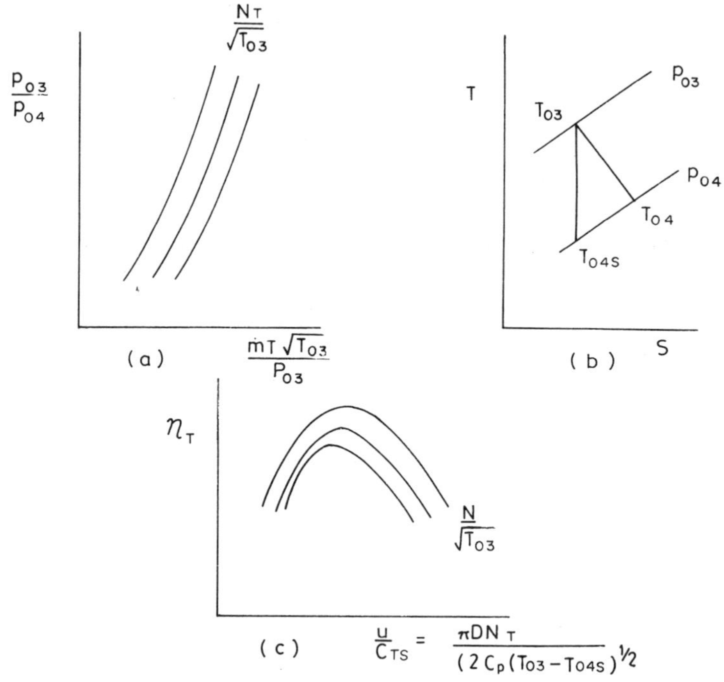

FIG. 10.36. Turbine characteristics.

From the air/fuel ratio the temperature rise (T_{03}-T_{02}) is evaluated from Fig. 10.37(b), the latter obtained from Fig. 10.11 and equation (10.15) for a given engine.

The compressor flow rate $\dot{m}_C = \dot{m}_a$ and the boost are known, so the compressor speed N_C can be determined from Fig. 10.37(c) and the power $\dot{W}_C$ from Fig. 10.37(d).

Due to the non-steady nature of the flow discussed earlier, the average turbine efficiency η_{TA} will be different from the value given in Fig. 10.36(c). If we let PF be the correlation factor such that

$$\eta_{TA} = PF\ \eta_T, \tag{10.78}$$

the turbine work $\dot{W}_T$ will be

$$\dot{W}_T = \eta_{TA}\eta_M\dot{W}_C, \tag{10.79}$$

where η_M is the mechanical efficiency of the turbocharger.

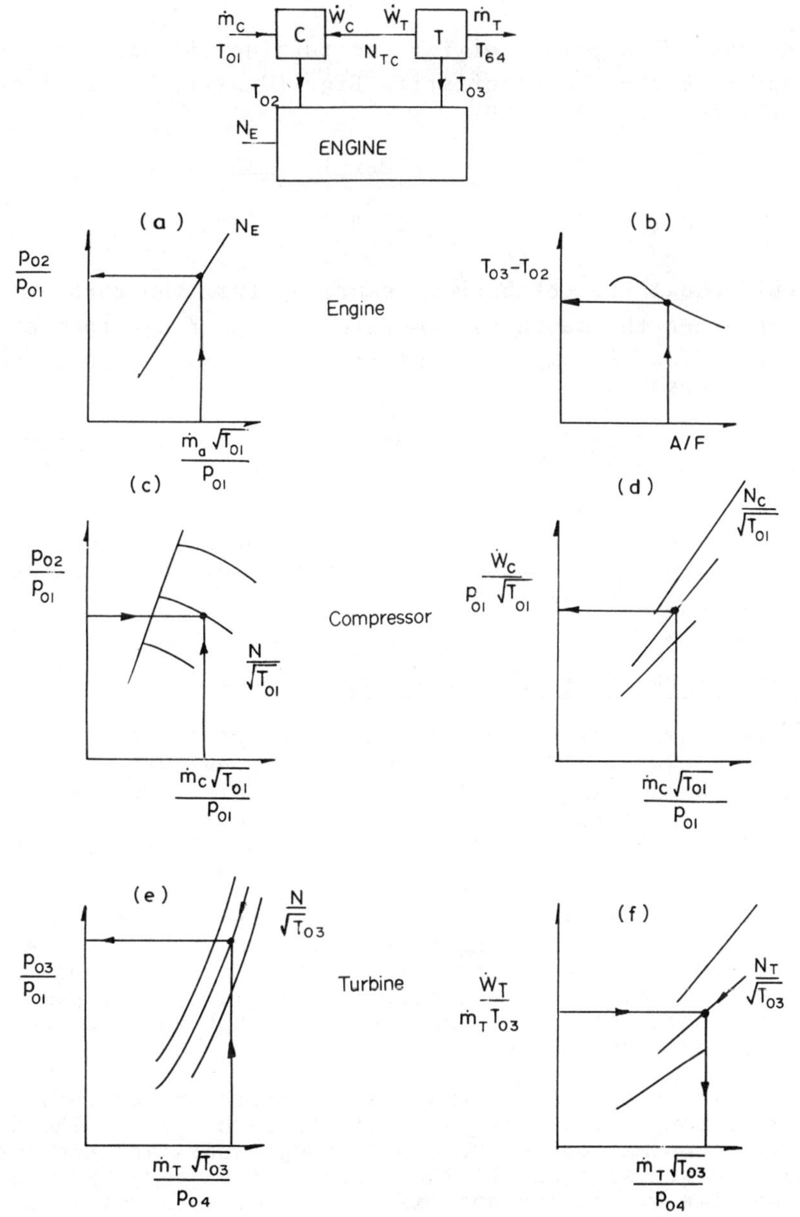

FIG. 10.37. Matching turbocharger/engine configuration.

Now

$$\dot{m}_T = \dot{m}_a + \dot{m}_f,$$

$$T_{03} = T_{02} + \left(T_{03}-T_{02}\right),$$

$$N_T = N_C.$$

Hence the mass flow parameter for the turbine $(\dot{m}_T \sqrt{T_{03}})/p_{04}$ can be determined from the characteristics Fig. 10.37(f). The turbine exhaust pressure p_{04} is then

$$p_{04} = \left(\frac{p_{04}}{\dot{m}_T\sqrt{T_{03}}}\right)\dot{m}_T\sqrt{T_{03}}. \tag{10.80}$$

This should equal the set back pressure p_b from the turbine exhaust. If $p_b = p_{04}$ then the match is completed, if $p_b \neq p_{04}$ then either another mass flow rate $\dot{m}_a$ or boost (p_{02}/p_{01}) is selected and the calculation repeated.

The pulse factor PF is dependent on the exhaust system design. The selection of a suitable valve will depend on test data.

Other methods for matching turbochargers to engines are based on a detailed calculation of the gas exchange process allowing for the non-steady flows. One method based on the method of characteristics is given by Benson.[9-11]

10.10 HIGH PRESSURE TURBOCHARGING

For very high engine outputs the boost pressures may be higher than can be achieved with conventional turbochargers, in these cases two-stage turbocharging may be used[12]. This involves two turbines and two compressors. A diagrammatic arrangement of such a system is shown in Fig. 10.38. In this system the engine exhausts into a high pressure turbine where the gas partially expands. It then passes to a low pressure turbine where the expansion is completed. The low pressure turbine drives a low pressure compressor and the high pressure turbine drives the high pressure compressor. The air enters the low pressure compressor and then passes through an intercooler. From the intercooler the air is further compressed in the high pressure compressor and passes through an aftercooler from whence it enters the engine.

Other high pressure systems use reciprocating or rotary displacement compressors with a single turbocharger. There are also many hybrid combinations of turbines, compressors and engines. Space precludes a discussion of all the variants that have been developed, but at the time of writing none is in general commercial use except for special applications.

10.11 SOME TURBOCHARGED ENGINE PERFORMANCE CHARACTERISTICS

At the beginning of this chapter we stated that the objective of supercharging was to increase the air quantity supplied to the engine in order to consume more fuel and hence increase the specific power output. We have described the basic thermodynamic and fluid

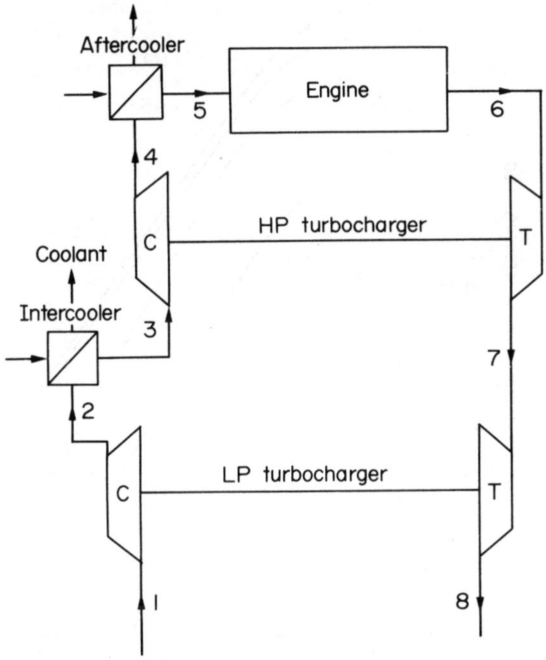

FIG. 10.38. Diagrammatic arrangement of two-stage series turbocharged engine. T, turbine; C, compressor.

mechanics considerations in order to successfully supercharge an engine. We shall conclude with a discussion of same performance characteristics of supercharged engines.

The degree of supercharging will be dependent not only on the satisfactory matching of the compressor, turbine and engine but also on the allowable maximum mechanical and thermal stresses. Designers normally specify certain limiting parameters for a given engine speed; these will be maximum cylinder pressure, maximum exhaust temperature and smoke limit. If the cylinder pressures are not measured the maximum boost might be specified.

In most cases of supercharged engines the performance characteristics are tailored to the engine application. We shall confine our discussion to relatively small engines (less than 10-litre capacity, four-stroke cycle), with application to automotive power units and electrical generating.

In Fig. 10.39 the brake mean effective pressure (b.m.e.p.) is shown over a 1200-2200 rev/min speed range for a four-cylinder, 7-litre, four-stroke engine to be used for automotive power. In these tests the supercharge air was provided from an external air supply. The increase in b.m.e.p. with decrease in engine speed is called the "torque back-up". The engine torque is directly proportional to the b.m.e.p. The higher the back-up the smaller the

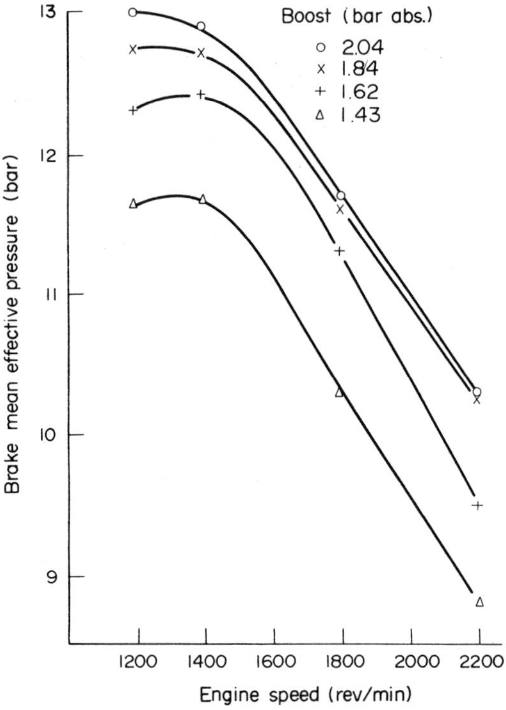

FIG. 10.39. Variation of b.m.e.p. with engine speed for constant boost. 7-litre engine, four-stroke cycle.

number of gears required for automotive drives. It is desirable to obtain these type of characteristics with the air supply obtained from a compressor driven by an exhaust turbine. In Fig. 10.40 the results are given for a comprehensive series of tests for this engine.

Since the engine is a four-cylinder four-stroke it is necessary, in order to avoid interference between cylinders, to group numbers 1 and 4 cylinders and numbers 2 and 3 cylinders into two separate exhaust pipes which are connected to separate entries to the turbine. The turbine then operates under partial admission at a lower efficiency than would be the case if a single entry was used without interference. It is possible by a special design to join the two exhaust pipes before entry to the turbine to avoid interference between the two groups of cylinders.(13) With this system a single entry can be achieved.

In the tests shown in Fig. 10.40 three sizes of radial inflow nozzle turbines were tested. The equivalent nozzle area is proportional to the A/R size which is specified for full admission. For partial admission the equivalent nozzle area is proportional to half the A/R size. The exhaust systems A and B were the conventional exhaust systems with partial admission. System A had a centre-mounted turbocharger and system B an end-mounted turbocharger.

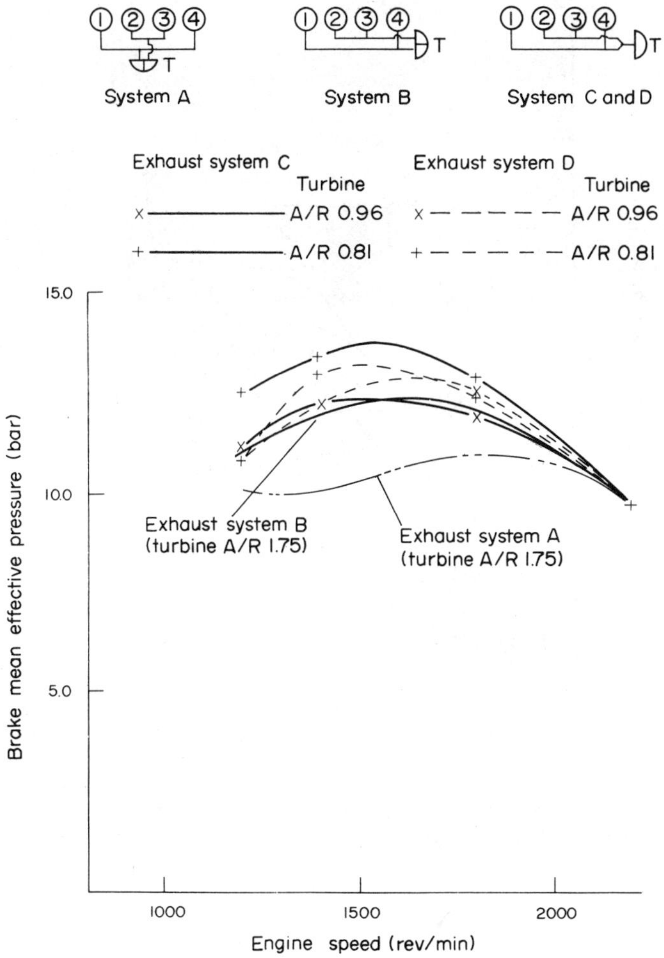

FIG. 10.40. Influence of turbocharger size and exhaust system on engine power. Four-cylinder, 7-litre engine, four-stroke cycle. All exhaust systems with end-mounted turbocharger except system A. System A centre-mounted turbocharger.

Systems C and D were the special systems with single-entry turbines. The difference between systems C and D lay in the geometrical proportions of the junction joining the two separate exhaust ducts ahead of the turbine. In all the tests the same compressor was fitted to the turbocharger.

The same engine test limits were specified for all configurations. These were fixed as maximum allowable boost at full speed, maximum smoke limit and exhaust temperature over the whole speed range. The results shown in Fig. 10.40 show significant differences for the torque back up and clearly illustrate the importance of the exhaust system design. A full discussion on the

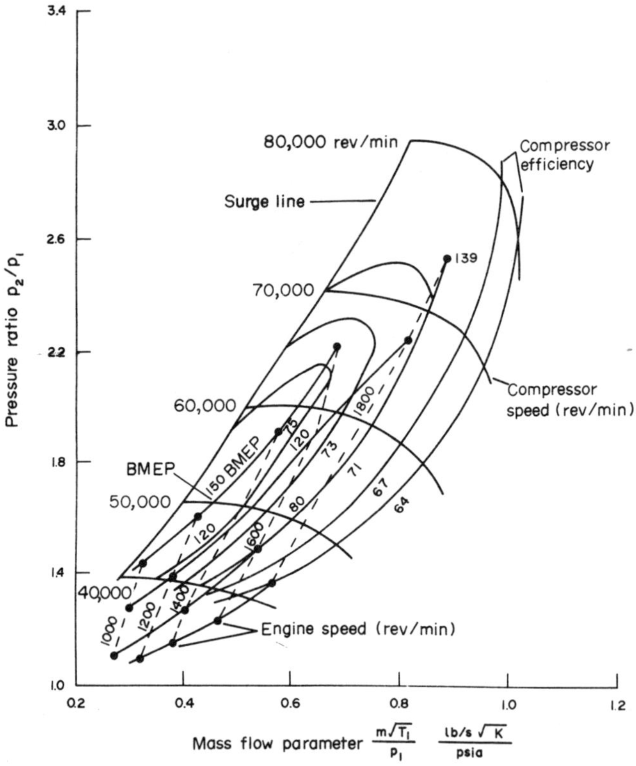

FIG. 10.41. Performance characteristics of turbocharged six-cylinder, four-stroke, 11.4-litre engine.

special exhaust systems is given in a paper by Benson.(13)

For electrical power generation different types of characteristics are required. A typical performance map for a small engine of 11.4 litres is shown in Fig. 10.41. In this figure we have the engine performance superimposed on the compressor characteristics. The latter comprise the conventional constant speed and efficiency lines. The engine characteristics are represented by lines of constant engine speed and torque (b.m.e.p.). You will see from these curves that at low engine speeds (1000 rev/min) and high torque (b.m.e.p.) the compressor is operating near the surge point, whilst at high engine speed and high torque the operating conditions are away from the surge line. The shape of the engine curves are due to the combustion characteristics of the engine. This is illustrated in Fig. 10.42 for an engine of similar design but less cylinders and slightly smaller bore. In Fig. 10.42(a) the engine test results are shown separately from the compressor. In this case the b.m.e.p. is shown for "constant" fuel pump setting. The term "constant" is in inverted commas because some slight adjustment had to be made to maintain a constant fuel rate. It will be seen that these curves have some resemblance to the torque back up

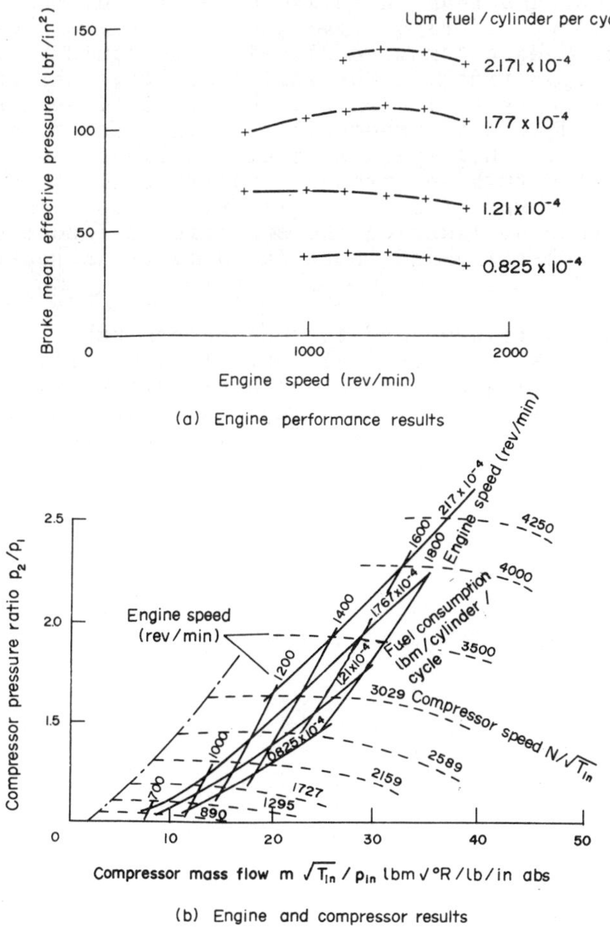

FIG. 10.42. Performance characteristics of turbocharged four-cylinder, four-stroke, 6.4-litre engine. (From R.S. Benson and P.C. Baruah.[11] By courtesy Society of Automotive Engineers.)

curves in Fig. 10.40 but are much flatter. Indeed we can see that constant fuel pump setting is almost equivalent to constant torque and the resultant performance map, Fig. 10.42(b), with lines of constant fuel rate, is similar to the torque (b.m.e.p.) lines in Fig. 10.41.

It should be emphasized that the particular form of the engine characteristic is dependent on the application, and that the engine designer attempts to achieve this by selecting the appropriate fuel injection system and turbocharger arrangement. The final trimming of the system is normally carried out on the test bed.

In recent years turbocharging has been increasingly applied to smaller-sized engines than those whose results are given in Figs. 10.39-10.41. In Fig. 10.43 the results are shown for a small 2.1-litre high speed automotive diesel which could be fitted in an automobile or light truck. The full lines give the engine performance naturally aspirated and the broken lines when the same engine was fitted with a turbocharger. In order to limit the maximum pressure at full speed some of the exhaust gases were bypassed round the turbine through a valve, called a wastegate. This produced the characteristic horizontal boost line above 2500 rev/min. In this application the specific fuel consumption was lower with turbocharging and there was a large increase in power output.

The reader is recommended to read papers published by the professional bodies and technical journals for performance characteristics of engines for other applications such as marine propulsion, pump drives, earth-moving equipment and so forth. Each application has a different type of engine performance characteristic.

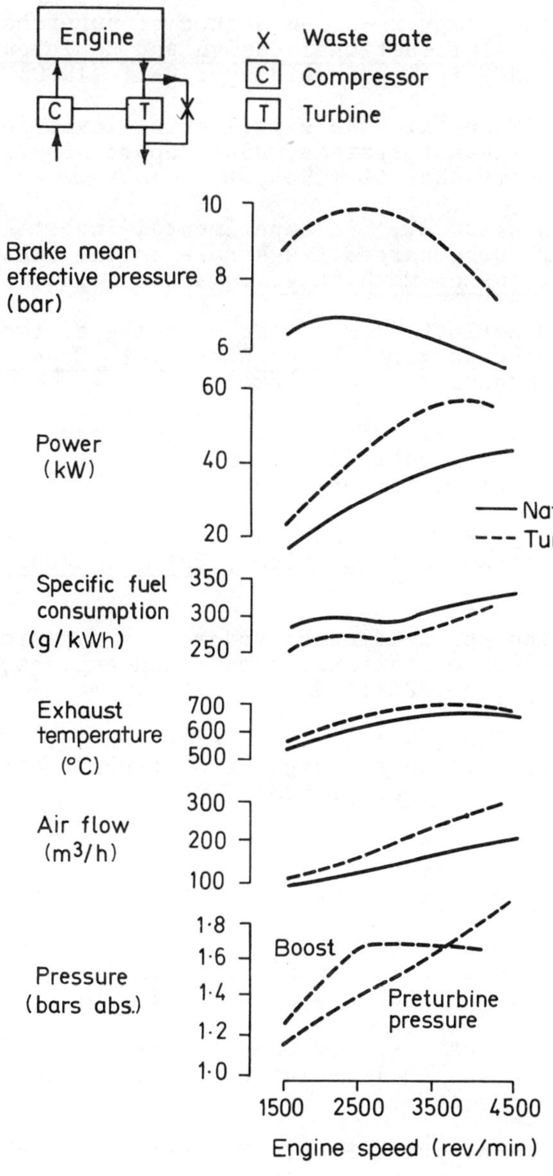

FIG. 10.43. Comparative performance of 2.1-litre diesel engine with and without turbocharger.

REFERENCES

1. Benson, R.S., The comprex—a new method of supercharging diesel engines, International Design and Equipment, Shipbuilding and Shipping Record, pp. 9-11 (1960).

2. Benson, R.S. and Wild, A., An experimental investigation into multi-cylinder exhaust systems, BSRA Report NS 92, Marine Engineering Report, No. 62 (1965).

3. Benson, R.S. and Wild, A., An experimental investigation into multi-cylinder turbocharged two-stroke engine exhaust systems, Proc. Instn. Mech. Engrs. Part 3J, 180, 290 (1965-6).

4. Benson, R.S. and Horlock, J.H., The matching of two-stroke engines and turbochargers, Proc. Sixth Int. Cong. Combustion Engines, Copenhagen, p. 464 (1962).

5. Smalley, D., An analysis of the two-stage series turbo-charging of a four-stroke diesel engine, University of Manchester, MSc dissertation, 1968.

6. Benson, R.S., Garg, R.D. and Woollatt, D., A numerical solution of unsteady flow problems, Int. J. Mech. Sci. 6 117 (1964).

7. Benson, R.S., The application of modern gas dynamic theories to exhaust systems of internal combustion engines, Trans. Lpool Engng. Soc. 76, 88 (1957).

8. Benson, R.S. and Woods, W.A., The energy content of exhaust pulses in a high pressure charged two-stroke cycle exhaust system, Int. J. Mech. Sci. 2, 231 (1960).

9. Benson, R.S., A computer program for calculating the performance of an internal combustion engine exhaust system, Proc. Instn. Mech. Engrs., Part 3L, 182, 91 (1967-8).

10. Benson, R.S., A comprehensive digital computer program to simulate a compression ignition engine including intake and exhaust systems, SAE, Paper 710173 (1971).

11. Benson, R.S. and Baruah, P.C., Some further tests on a computer program to simulate internal combustion engines, SAE, Paper 730667 (1973).

12. Benson, R.S. and Svetnicka, F.V., Two-stage turbocharging of diesel engines: a matching procedure and an experimental investigation, SAE, Paper 740740 (1974).

13. Benson, R.S., and Alexander, G.I., The application of pulse converters to automotive four-stroke cycle engines: Part II, Optimisation of pulse converter-turbocharger combination, SAE, Paper 770034 (1977).

14. Daneshyar, H., One-dimensional Compressible Flow, Pergamon Press, Oxford, 1976.

Appendix II

A. THERMODYNAMIC PROPERTIES OF MIXTURES

The thermodynamic properties of mixtures may be computed by fairly simple functions. The specific enthalpy $h_i(T)$ and internal energy $e_i(T)$ are given in Chapter 2 (expressions (2.104) and (2.105)) as

$$h_i(T) = R_{mol}\left(\sum_{j=1}^{j=5} u_{i,j}\ T^j\right), \tag{II.1}$$

$$e_i(T) = R_{mol}\left(\left(\sum_{j=1}^{j=5} u_{i,j}\ T^j\right) - T\right). \tag{II.2}$$

The coefficients $u_{i,j}$ are given in Table 2.1 (Vol.I, page 40). For a mixture of gases, species i = 1 to N, the total internal energy E(T) is

$$E(T) = R_{mol}\sum_{i=1}^{i=N} w_i\left(\left(\sum_{j=1}^{j=5} u_{i,j}\ T^j\right) - T\right), \tag{II.3}$$

where w_i is the number of mols of gas i.

In Table 2.1 the following subscripts are given for the species:

Species	CO_2	CO	H_2O	H_2	O_2	N_2	C_nH_m
Subscript (i)	1	2	3	4	5	6	7

The internal energy of the mixture E(T) can be expressed in terms of the specific internal energy of the mixture e(T) and the total number of mols w_m in the mixture given by

$$w_m = \sum_{i=1}^{i=N} w_i$$

whence $$E(T) = w_m e(T).$$

Now the specific heat at constant volume $C_v(T)$ for the mixture is

$$C_v(T) = \frac{\partial e(T)}{\partial T}.$$

Hence the specific heat is

$$C_v(T) = \frac{1}{w_m}\frac{\partial E(T)}{\partial T} = \frac{E'(T)}{w_m}.$$

Differentiating (II.3) with respect to T we have

$$w_m C_v(T) = R_{mol} \sum_{i=1}^{i=N} w_i \left(\left(\sum_{j=1}^{j=5} j u_{i,j} \, T^{j-1} \right) \right). \tag{II.4}$$

Equations (II.3) and (II.4) are general expressions for internal energy and specific heat for the mixture.

To calculate the equilibrium constants K_p we require the Gibbs functions for the species. The expressions for the Gibbs function is, from (2.145),

$$G_i(T) = \frac{g_i(T)}{R_{mol}T} = u_{i,1}(1-\ell n\ T) - \sum_{j=2}^{j=5} \frac{u_{i,j}}{j-1} T^{j-1} - u_{i,6}$$

or

$$G_i(T) = u_{i,1}\ (1-\ell n\ T) - V(T)_i - u_{i,6}, \tag{II.5}$$

where

$$V(T)_i = \sum_{j=2}^{j=5} \frac{u_{i,j}}{j-1} T^{j-1}. \tag{II.6}$$

For example, for the two reactions (9.18) and (9.19) we can evaluate K_p with the aid of (II.5) and (II.6).

For reaction (9.18):

$$\begin{aligned} \ell n\left(K_{p1}\right) = & \left(u_{2,1} + u_{3,1} - u_{4,1} - u_{1,1}\right)(1-\ell n\ T) \\ & -\left(V(T)_2 + V(T)_3 - V(T)_4 - V(T)_1\right) \\ & -\left(u_{2,6} + u_{3,6} - u_{4,6} - u_{1,6}\right) + \frac{0.4047 \times 10^8}{R_{mol}T} \end{aligned} \tag{II.7}$$

For reaction (9.19):

$$\begin{aligned} \ell n\left(K_{p2}\right) = & \left(u_{2,1} + 0.5u_{5,1} - u_{1,1}\right)(1-\ell n\ T) \\ & -\left(V(T)_2 + 0.5V(T)_5 - V(T)_1\right) \\ & -\left(u_{2,6} + 0.5u_{5,6} - u_{1,6}\right) + \frac{2.7969 \times 10^8}{R_{mol}T} \end{aligned} \tag{II.8}$$

Numerical Solutions

Simple algorithms can be developed for (II.3) and (II.4). We shall define:

$$f(u)_i = \sum_{j=1}^{j=1} u_{i,j} y^{j-1}, \tag{II.9}$$

$$f(udy)_i = \sum_{j=1}^{j=5} ju_{i,j}\, y^{j-1}, \tag{II.10}$$

$$f(tu) = \sum_{i=I1}^{i=I2} w_i(f(u)_i - 1), \tag{II.11}$$

$$f(tudy) = \sum_{i=I1}^{i=I2} w_i(f(udy)_i - 1). \tag{II.12}$$

It follows from (II.9) to (II.12) that (II.3) and (II.4) can be written as

$$E(T) = R_{mol} T\, f(tu)_{y=T}, \tag{II.13}$$

$$w_m C_v(T) = R_{mol}\, f(tudy)_{y=T}. \tag{II.14}$$

Example

To illustrate the method of solution we shall take the reactions given in the spark ignition cycle calculations, Chapter 9.

The reactants comprise O_2, N_2 and C_nH_m.

Let w_5 = number of mols of O_2, subscript i=5,

w_6 = number of mols of N_2, subscript i=6,

w_7 = number of mols of C_nH_m, subscript i=7,

then the limits are I1=5 and I2=7.

If the reactant temperature is T_R, then:

$$y = T_R,$$

$$f(tu)_{y=T_R} = \sum_{i=5}^{i=7} w_i(f(u)_i - 1)_{y=T_R}$$

$$f(tudy)_{y=T_R} = \sum_{i=5}^{i=7} w_i(f(udy)_i - 1)_{y=T_R},$$

$$w_R = \sum_{i=5}^{i=7} w_i.$$

The internal energy and specific heats are then:

$$E_R(T_R) = R_{mol} T_R f(tu)_{y=T_R},$$

$$C_v(T_R) = \frac{R_{mol}}{W_R} f(tudy)_{y=T_R}.$$

The products comprise CO_2, CO, H_2O, H_2, O_2 and N_2. We use the notation:

i	Species		
1	CO_2	4	H_2
2	CO	5	O_2
3	H_2O	6	N_2

then I1 = 1 and I2 = 6.

The temperature of the products is T_P and $y = T_P$, then:

$$f(tu)_{y=T_P} = \sum_{i=1}^{i=6} w_i(f(u)_i - 1)_{y=T_P},$$

$$f(tudy)_{y=T_P} = \sum_{i=1}^{i=6} w_i(f(udy)_i - 1)_{y=T_P},$$

$$w_P = \sum_{i=1}^{i=6} w_1.$$

The internal energy and specific heats are then:

$$E_P(T_P) = R_{mol} T_P f(tu)_{y=T_P},$$

$$C_v(T_P) = \frac{R_{mol}}{w_P} f(tudy)_{y=T_P}.$$

The algorithms (II.9), (II.10) and (II.6) are programmed in a subroutine FNU and the algorithms (II.11) and (II.12) are programmed in subroutine FNUT in Table II.1.

In FNU the variables are:

F1 = $f(u)_i$ equation (II.9),

F2 = $f(udy)_i$ equation (II.10),

F3 = $V(T)_i$ equation (II.6),

In FNUT the variables are:

F1 = f(tu),

F2 = f(tudy),

$$F3 = w_m = \sum_{i=1}^{i=N} w_i.$$

The polynomial coefficients are U(7,7).

TABLE II.1 Subroutines to Calculate Thermodynamic Properties of Mixtures.

```
      SUBROUTINE FNUT(F1,F2,F3,I1,I2,X,FU,FUDY)
      DIMENSION FU(7),FUDY(7),X(7)
      F1=0.0
      F2=0.0
      F3=0.0
      DO 1 I=I1,I2
      F1=F1+X(I)*(FU(I)-1.0)
      F2=F2+X(I)*(FUDY(I)-1.0)
    1 F3=F3+X(I)
      RETURN
      END
```

```
      SUBROUTINE FNU(F1,F2,F3,I,U,Y,MARK)
      DIMENSION U(7,7)
      F1=0.0
      F2=0.0
      F3=0.0
      DO 1 J=1,5
      S=FLOAT(J)
      F1=F1+U(I,J)*Y**(S-1.0)
      F2=F2+S*U(I,J)*Y**(S-1.0)
      IF(MARK.EQ.0)GO TO 1
      IF(J.EQ.1)GO TO 1
      F3=F3+(U(I,J)*Y**(S-1.0))/(S-1.0)
    1 CONTINUE
      RETURN
      END
```

B. DUAL COMBUSTION CYCLE PROGRAM

The theory outlined in Chapter 8 for the ideal cycle has been programmed in ASA FORTRAN IV. The listing is given in Table II.2. The listing of the two subroutines, (Table II.1) is included. The data required are:

Card No.	Name	Description
1	NDSET	Number of sets of calculations
2	U(5,L), L=1,7	Seven polynomial coefficients for the fuel (see Table 2.1)
3	D	Cylinder bore (m)
	S	Cylinder stroke (m)
	CR	Compression ratio
4	AFR	Air/fuel ratio
	CA	Percentage carbon by weight in fuel
	HA	Percentage hydrogen by weight in fuel
	QVS	Heat of reaction (J/kg). (This is converted to J/kg-mol in program)
	XF	Fraction of fuel burnt in constant volume phase (0 to 1.0)
5	PIN	Pressure at commencement of compression stroke (N/m^2)
	TIN	Temperature at commencement of compression stroke (K)
	PO	Reference pressure (10132.5 N/m^2)
	TS	Reference temperature (288 K)
	REV	Working cycles per second
6	ND	Number of volume divisions for compression and expansion stroke not to exceed 50
7	ACCUR1	Accuracy for solutions of Newton-Raphson equations set to 0.1

The formats for cards 2, 3, 4, 5, is 8F 10.8.

For card 7 2F 11.6 and for cards 1 and 6 8I 10.

TABLE II.2. Fortran Listing for Dual Combustion Cycle.

```
      PROGRAM DUAL(INPUT,OUTPUT,TAPE5=INPUT,TAPE6=OUTPUT)
C DUAL COMBUSTION CYCLE ANALYSIS PROGRAM
      DIMENSION A(7),B(7),FU(7),FUDT(7),FV(7),P(2,50),T(2,50),
     1U(5,7),WORK(2,50),V(2,50),ENERGY(2,50),PBARA(50)
      READ(5,102) NDSET
C READ POLYNOMIAL COEFFICIENTS FOR FUEL
      READ(5,106) (U(5,L),L=1,7)
  106 FORMAT(E14.5)
      DO 9999 NX=1,NDSET
      READ(5,101)D,S,CR
      READ(5,101)AFR,CA,HA,QVS,XF
      READ(5,101)PIN,TIN,PO,TS,REV
      READ(5,102)ND,NFRST
      READ(5,104) ACCUR1
  101 FORMAT(8F10.0)
  102 FORMAT(8I10)
  104 FORMAT(2F11.6)
      WRITE(6,108)
  108 FORMAT(1H1)
      DO 2 II=1,7
      A(II)=0.0
    2 B(II)=0.0
C PRINT INPUT DATA
      WRITE(6,109)
  109 FORMAT(1H ,//10X,38HDUAL COMBUSTION CYCLE ANALYSIS PROGRAM/
     110X,38H--------------------------------------///
     110X,10HINPUT DATA/10X,10H----------/)
      WRITE(6,113)D,S,CR,AFR
  113 FORMAT(10X,11HCYL DIA (M),F8.3//10X,10HSTROKE (M),F8.3//
     110X,10HCOMP RATIO,F7.2//10X,14HAIR FUEL RATIO,F7.2/)
      WRITE(6,111)CA,HA,QVS,XF,PIN,TIN,PO,TS,REV
  111 FORMAT(10X,20HCARBON ATOMS IN FUEL,F8.2//10X,
     122HHYDROGEN ATOMS IN FUEL,F8.2//10X,
     123HHEAT OF REACTION (J/KG),F15.1//10X,
     152HFRACTION OF FUEL BURNT IN CONSTANT VOLUME COMBUSTION,
     1F8.4//10X,22HTRAPPED PRESS (N/M**2),F15.3//10X,
     116HTRAPPED TEMP (K),F10.1//10X,
     118HREF PRESS (N/M**2),F15.3//10X,
     112HREF TEMP (K),F10.1//10X,
     118HENGINE SPEED (RPS),F10.1)
C      PREPARE THERMODYNAMIC DATA U(I,J)
C      I=SPECIES NO,J=COEFFICIENT
C      SPECIES NUMBER
C      CO2=1,H2O=2,O2=3,N2=4,CNHM=5
C      COEFFICIENTS
C      1=A,2=B,3=C,4=D,5=E,6=K,7=HO
C CO2
      U(1,1)=3.09590E00
      U(1,2)=2.73114E-03
      U(1,3)=-7.88542E-07
      U(1,4)=8.66002E-11
      U(1,5)=0.00000E00
```

```
      U(1,6)=6.58393E00
      U(1,7)=-3.93640E08
C H2O
      U(2,1)=3.74292E00
      U(2,2)=5.65590E-04
      U(2,3)=4.95240E-08
      U(2,4)=-1.81802E-11
      U(2,5)=0.00000E00
      U(2,6)=9.65140E-01
      U(2,7)=-2.39225E08
C O2
      U(3,1)=3.25304E00
      U(3,2)=6.52350E-04
      U(3,3)=-1.49524E-07
      U(3,4)=1.53897E-11
      U(3,5)=0.00000E00
      U(3,6)=5.71243E00
      U(3,7)=0.00000E00
C N2
      U(4,1)=3.34435E00
      U(4,2)=2.94260E-04
      U(4,3)=1.95300E-09
      U(4,4)=-6.57470E-12
      U(4,5)=0.00000E00
      U(4,6)=3.75863E00
      U(4,7)=0.00000E00
      WORKT=0.0
      RMOL=8314.3
      PI=3.1415927
C SET UP INPUT DATA
      VS=PI*S*(D/2.0)**2
      VC=VS/CR
      V1=VS+VC
      ND=ND-1
      DV=VS/FLOAT(ND)
      P1=PIN
      T1=TIN
C CALCULATE MOLS OF FUEL W
      SOX=CA+0.25*HA
      WF=12.0*CA+HA
C STOICHIOMETRIC AIR FUEL RATIO BY WEIGHT
      AFST=4.7619*SOX*28.96/WF
      PHI=AFST/AFR
      WM1=4.76*(CA+0.25*HA)/PHI
      W=(P1*V1)/(WM1*RMOL*T1)
      WRITE(6,164) AFST,PHI
  164 FORMAT(1H ,//10X,29HAIR FUEL RATIO STOICHIOMETRIC,F8.3//
     1   10X,20HFUEL AIR EQUIVALENCE,F10.3)
      WRITE(6,115)
  115 FORMAT(1H ,3X,4HSTEP,3X,9HVOL(M**3),2X,10HPRESS(BAR),
     11X,7HTEMP(K),4X,9HENERGY(J)/)
C CONVERT HEAT OF REACTION FROM J/KG TO J/KGMOL
      QVS=QVS*WF
C TOTAL MOLES(KGMOL) OF O2, N2, CNHM AT TRAPPED CONDITION
      DO 4 I=1,5
    4 B(I)=0.0
      B(3)=W*SOX/PHI
```

```
      B(4)=3.76*B(3)
      P(1,1)=P1
      T(1,1)=T1
      V(1,1)=V1
      N=1
      PBAR=P(1,1)*1.OE-5
      PBARA(1)=PBAR
      NC=ND+1
C INTERNAL ENERGY AT TRAPPED STATE FOR AIR
      DO 10 I=3,4
   10 CALL FNU(FU(I),FUDT(I),FV(I),I,U,T1,0)
      CALL FNUT(FUT,FUDTT,WM,3,4,B,FU,FUDT)
      ET1=RMOL*T1*FUT
      ENERGY(1,1)=-W*QVS
      CV=RMOL*FUDTT/WM
C START CALCULATION FOR COMPRESSION STROKE
      DO 14 N=2,NC
C ESTIMATE T2 AND P2
      V2=V1-DV
      T2=T1*(V1/V2)**(RMOL/CV)
      P2=P1*(V1/V2)*(T2/T1)
C INTERNAL ENERGY AT STATE 2
   13 DO 11 I=3,4
   11 CALL FNU(FU(I),FUDT(I),FV(I),I,U,T2,0)
      CALL FNUT(FUT,FUDTT,WM,3,4,B,FU,FUDT)
      ET2=RMOL*T2*FUT
      CV=RMOL*FUDTT/WM
C WORK
      DW=0.5*(P2+P1)*(V2-V1)
C FIRST LAW
      FE=(ET2-ET1)+DW
      ERROR=FE/(WM*CV)
      IF(ABS(ERROR).LT.ACCUR1) GO TO 12
C IF FIRST LAW NOT SATISFIED ESTIMATE NEW
C VALUE OF T2 AND P2 AND RETURN TO STATEMENT 13
      T2=T2-ERROR
      P2=P1*(V1/V2)*(T2/T1)
      GO TO 13
C FIRST LAW SATISFIED STORE VALUES OF PC,TC,DW AND RESET T1,P1,ET1
   12 T(1,N)=T2
      P(1,N)=P2
      V(1,N)=V2
      ENERGY(1,N)=ENERGY(1,N-1)+ET2-ET1
      WORK(1,N)=DW
      ET1=ET2
      T1=T2
      P1=P2
      V1=V2
      WORKT=WORKT+DW
      PBAR=P(1,N)*1.OE-5
      PBARA(N)=PBAR
   14 CONTINUE
C END OF COMPRESSION STROKE
C COMBUSTION AT CONSTANT VOLUME
C SET UP INITIAL CONDITIONS
      TR=T2
      PR=P2
```

```
      VC=V2
      B(5)=W*XF
C ENERGY LEVEL OF REACTANTS AT TR
      DO 20 I=3,5
   20 CALL FNU(FU(I),FUDT(I),FV(I),I,U,TS,0)
      CALL FNUT(FUTS,FUDTS,WMR,3,5,B,FU,FUDT)
      DO 22 I=3,5
   22 CALL FNU(FU(I),FUDT(I),FV(I),I,U,TR,0)
      CALL FNUT(FUTR,FUDTR,WM,3,5,B,FU,FUDT)
      E1=RMOL*TR*FUTR-RMOL*TS*FUTS
      CV1=RMOL*FUDTR/WM
C PRINT COMPRESSION STROKE
      DO 24 N=1,NC
      ENERGY(1,N)=ENERGY(1,N)
   24 WRITE(6,160)N,V(1,N),PBARA(N),T(1,N),ENERGY(1,N)
C PRODUCTS
      A(1)=XF*W*CA
      A(2)=XF*W*HA/2.
      A(3)=B(3)-XF*W*(CA+0.25*HA)
      A(4)=B(4)
      A(5)=0
C ESTIMATE T2
      T2=TR-XF*W*QVS/(WM*CV1)
C SOLVE FOR NEW VALUE OF T2 BY ENERGY BALANCE
      DO 44 I=1,5
   44 CALL FNU(FU(I),FUDT(I),FV(I),I,U,TS,0)
      CALL FNUT(FUTS,FUDTS,WM,1,5,A,FU,FUDT)
   47 DO 46 I=1,5
   46 CALL FNU(FU(I),FUDT(I),FV(I),I,U,T2,0)
      CALL FNUT(FUTP,FUDTP,WM,1,5,A,FU,FUDT)
      E2=RMOL*T2*FUTP-RMOL*TS*FUTS
      FE=E2-E1+XF*W*QVS
      CV=RMOL*FUDTP/WM
      ERROR=FE/(WM*CV)
      IF(ABS(ERROR).LT.ACCUR1)GO TO 50
      T2=T2-ERROR
      GO TO 47
C CONDITIONS AT END OF CONSTANT VOLUME COMBUSTION
   50 P2=(WM/WMR)*(T2/TR)*PR
      P(2,1)=P2
      T(2,1)=T2
      V(2,1)=V2
      ENERGY(2,1)=ENERGY(1,50)
      WORK(2,1)=0.0
      N=1
      PBAR=P2*1.0E-5
      WRITE(6,160)N,V2,PBAR,T2,ENERGY(2,1)
C COMBUSTION AT CONSTANT PRESSURE REACTANTS
      DO 30 I=1,4
   30 B(I)=A(I)
      B(5)=(1.-XF)*W
C PRODUCTS
      A(1)=W*CA
      A(2)=W*HA/2.
      A(3)=W*(CA+.25*HA)/PHI
      A(4)=B(4)
      A(5)=0.0
```

```
      WMC=WM+B(5)
C SOLVE FOR NEW VALUE OF T2 BY ENERGY BALANCE
      T1=T2
      P1=P2
      V1=V2
      E1=E2
      T2=T1-(1.-XF)*W*QVS/(WM*CV)
      DO 52 I=1,5
   52 CALL FNU(FU(I),FUDT(I),FV(I),I,U,TS,0)
      CALL FNUT(FUTS,FUDTS,WM,1,5,A,FU,FUDT)
  500 DO 182 I=1,5
  182 CALL FNU(FU(I),FUDT(I),FV(I),I,U,T2,0)
      CALL FNUT(FUTP,FUDTP,WM,1,5,A,FU,FUDT)
      E2=RMOL*T2*FUTP-RMOL*TS*FUTS
      V2=(WM*T2*V1)/(WMC*T1)
      DW=P2*(V2-V1)
      FE=E2-E1+(1.-XF)*W*QVS+DW
      CV=RMOL*FUDTP/WM
      ERROR=FE/(CV*WM)
      T2=T2-ERROR
      IF(ABS(ERROR).LT.ACCUR1) GO TO 501
      GO TO 500
C END OF COMBUSTION AT CONSTANT PRESSURE
  501 P(2,2)=P2
      T(2,2)=T2
      ENERGY(2,2)=ENERGY(2,1)+E2-E1+(1.0-XF)*W*QVS
      WORKT=WORKT+DW
      V(2,2)=V2
      VSE=VS+VC-V2
      DV-VSE/FLOAT(ND-1)
      N=2
      PBAR=P2*1.0E-5
      WRITE(6,160)N,V2,PBAR,T2,ENERGY(2,2)
C EXPANSION STROKE
      DO 70 N=3,NC
      T1=T2
      P1=P2
      V1=V2
      V2=V1+DV
C ESTIMATE T2 AND P2
      T2=T1*(V1/V2)**(RMOL/CV)
      P2=P1*(V1/V2)*(T2/T1)
C SOLVE FOR NEW VALUE OF T2 BY ENERGY BALANCE
      E1=E2
      DO 71 I=1,5
   71 CALL FNU(FU(I),FUDT(I),FV(I),I,U,TS,0)
      CALL FNUT(FUTS,FUDTS,WM,1,5,A,FU,FUDT)
  110 DO 112 I=1,5
  112 CALL FNU(FU(I),FUDT(I),FV(I),I,U,T2,0)
      CALL FNUT(FUT,FUDTT,WM2,1,5,A,FU,FUDT)
      E2=RMOL*T2*FUT-RMOL*TS*FUTS
      DW=0.5*(P1+P2)*(V2-V1)
      DE=(E2-E1)
      FE=DE+DW
      CV=RMOL*FUDTT/WM2
```

```
      ERROR=FE/(WM2*CV)
      IF(ABS(ERROR).LT.ACCUR1)GO TO 200
      T2=T2-ERROR
      P2=P1*(V1/V2)*(T2/T1)
      GO TO 110
C CONDITION AT END OF VOLUME STEP
  200 P(2,N)=P2
      T(2,N)=T2
      V(2,N)=V2
      ENERGY(2,N)=ENERGY(2,N-1)+E2-E1
      WORK(2,N)=DW
      WORKT=WORKT+DW
      PBAR=P(2,N)*1.0E-5
      WRITE(6,160)N,V(2,N),PBAR,T(2,N),ENERGY(2,N)
  160 FORMAT(1H ,3X,I2,2X,E12.4,3X,F8.3,2X,F6.1,2X,E12.5)
   70 CONTINUE
      EFFTH=100.0*WORKT/(-QVS*W)
      POWER=WORKT*REV/2000.0
      PMIP=WORKT*1.0E-5/VS
      WRITE(6,162)PMIP,POWER,EFFTH
  162 FORMAT(1H0,14X,12HI.M.E.P(BAR),F10.3,5X,
     120HPOWER(4 STROKE) (KW),
     1F10.3,5X,18HTHERMAL EFFICIENCY,F7.2)
 9999 CONTINUE
      STOP
      END


      SUBROUTINE FNU(F1,F2,F3,I,U,Y,MARK)
      DIMENSION U(5,7)
      F1=0.0
      F2=0.0
      F3=0.0
      DO 1 J=1,5
      S=FLOAT(J)
      F1=F1+U(I,J)*Y**(S-1.0)
      F2=F2+S*U(I,J)*Y**(S-1.0)
      IF(MARK.EQ.0)GO TO 1
      IF(J.EQ.1)GO TO 1
      F3=F3+(U(I,J)*Y**(S-1.0))/(S-1.0)
    1 CONTINUE
      RETURN
      END


      SUBROUTINE FNUT(F1,F2,F3,I1,I2,X,FU,FUDY)
      DIMENSION FU(7),FUDY(7),X(7)
      F1=0.0
      F2=0.0
      F3=0.0
      DO 1 I=I1,I2
      F1=F1+X(I)*(FU(I)-1.0)
      F2=F2+X(I)*(FUDY(I)-1.0)
    1 F3=F3+X(I)
      RETURN
      END
```

C. OTTO CYCLE PROGRAM

The theory outlined in Chapter 9 for the ideal cycle has been programmed in ASA FORTRAN IV. The listing is given in Table II.3. In addition to the listing, the two subroutines (Table II.1), are to be added. The data required are:

Card No.	Name	Description
1	NDSET	Number of sets of calculations
2	U(7,L), L=1,7	Seven polynomial coefficients for the fuel (see Table 2.1)
3	D	Cylinder bore (m)
	S	Cylinder stroke (m)
	CR	Compression ratio
4	AFR	Air/fuel ratio
	CA	Percentage carbon by weight in fuel
	HA	Percentage hydrogen by weight in fuel
	QVS	Heat of reaction (J/kg) (This is converted to J/kg-mol in program)
5	PIN	Pressure at commencement of compression stroke (N/m^2)
	TIN	Temperature at commencement of compression stroke (K)
	PO	Reference pressure (10132.5 N/m^2)
	TS	Reference temperature (288 K)
	REV	Working cycles per second
6	ND	Number of volume divisions for compression and expansion stroke: not to exceed 50
	NDFST	Number of volume increments in expansion stroke at which mixture can be considered to have frozen
7	ACCUR1	Accuracy for solution of equations type (9.14): set to 0.1.
	ACCUR2	Accuracy for solution of equation (9.28)

The format for cards 2, 3, 4, 5 is 8F 10.8, and for card 7 (2F 11.6).

For cards 1 and 6 the format is 8I 10.

TABLE II.3. Fortran Listing for Otto Cycle.

```
      PROGRAM OTTO(INPUT,OUTPUT,TAPE5=INPUT,TAPE6=OUTPUT)
C   OTTO CYCLE ANALYSIS PROGRAM
      DIMENSION A(7),B(7),FU(7),FUDT(7),FV(7),P(2,50),T(2,50),
     1U(7,7),WORK(2,50),V(2,50),VAF(7),ENERGY(2,50),
     1PBARA(50),VAFA(50,7)
      READ(5,102) NDSET
C READ POLYNOMIAL COEFFICIENTS FOR FUEL
      READ(5,106) (U(7,L),L=1,7)
  106 FORMAT(E14.5)
      DO 9999 NX=1,NDSET
      READ(5,101)D,S,CR
      READ(5,101)AFR,CA,HA,QVS
      READ(5,101)PIN,TIN,PO,TS,REV
      READ(5,102)ND,NFRST
      READ(5,104) ACCUR1,ACCUR2
  101 FORMAT(8F10.0)
  102 FORMAT(8I10)
  104 FORMAT(2F11.6)
      WRITE(6,108)
  108 FORMAT(1H1)
      DO 2 II=1,7
      A(II)=0.0
    2 B(II)=0.0
C PRINT INPUT DATA
      WRITE(6,109)
  109 FORMAT(1H ,//10X,27HOTTO CYCLE ANALYSIS PROGRAM/10X,
     127H---------------------------///10X,
     110HINPUT DATA/10X,10H----------/)
      WRITE(6,113)D,S,CR,AFR
  113 FORMAT(10X,11HCYL DIA (M),F8.3//10X,10HSTROKE (M),F8.3//
     110X,10HCOMP RATIO,F7.2//10X,14HAIR FUEL RATIO,F7.2/)
      WRITE(6,111)CA,HA,QVS,PIN,TIN,PO,TS,REV
  111 FORMAT(10X,20HCARBON ATOMS IN FUEL,F8.2//10X,
     122HHYDROGEN ATOMS IN FUEL,F8.2//10X,
     123HHEAT OF REACTION (J/KG),F15.1//10X,
     122HTRAPPED PRESS (N/M**2),F15.3//10X,
     116HTRAPPED TEMP (K),F10.1//10X,
     118HREF PRESS (N/M**2),F15.3//10X,
     112HREF TEMP (K),F10.1//10X,
     118HENGINE SPEED (RPS),F10.1)
C      PREPARE THERMODYNAMIC DATA U(I,J)
C      I=SPECIES NO,J=COEFFICIENT
C      SPECIES NUMBER
C      CO2=1,CO=2,H2O=3,H2=4,O2=5,N2=6,CNHM=7
C      COEFFICIENTS
C      1=A,2=B,3=C,4=D,5=E,6=K,7=HO
C CO2
      U(1,1)=3.09590E00
      U(1,2)=2.73114E-03
      U(1,3)=-7.88542E-07
      U(1,4)=8.66002E-11
      U(1,5)=0.00000E00
```

```
      U(1,6)=6.58393E00
      U(1,7)=-3.93640E08
C CO
      U(2,1)=3.31700E00
      U(2,2)=3.76970E-04
      U(2,3)=-3.22080E-08
      U(2,4)=-2.19450E-12
      U(2,5)=0.00000E00
      U(2,6)=4.63284E00
      U(2,7)=-1.13950E08
C H2O
      U(3,1)=3.74292E00
      U(3,2)=5.65590E-04
      U(3,3)=4.95240E-08
      U(3,4)=-1.81802E-11
      U(3,5)=0.00000E00
      U(3,6)=9.65140E-01
      U(3,7)=-2.39225E08
C H2
      U(4,1)=3.43328E00
      U(4,2)=-8.18100E-06
      U(4,3)=9.66990E-08
      U(4,4)=-1.44392E-11
      U(4,5)=0.00000E00
      U(4,6)=-3.84470E00
      U(4,7)=0.00000E00
C O2
      U(5,1)=3.25304E00
      U(5,2)=6.52350E-04
      U(5,3)=-1.49524E-07
      U(5,4)=1.53897E-11
      U(5,5)=0.00000E00
      U(5,6)=5.71243E00
      U(5,7)=0.00000E00
C N2
      U(6,1)=3.34435E00
      U(6,2)=2.94260E-04
      U(6,3)=1.95300E-09
      U(6,4)=-6.57470E-12
      U(6,5)=0.00000E00
      U(6,6)=3.75863E00
      U(6,7)=0.00000E00
      WORKT=0.0
      RMOL=8314.3
      PI=3.1415927
C SET UP INPUT DATA
      VS=PI*S*(D/2.0)**2
      VC=VS/CR
      V1=VS+VC
      ND=ND-1
      DV=VS/FLOAT(ND)
      P1=PIN
      T1=TIN
C CALCULATE MOLS OF FUEL W
      SOX=CA+0.25*HA
      WF=12.0*CA+HA
```

```
C STOICHIOMETRIC AIR FUEL RATIO BY WEIGHT
      AFST=4.7619*SOX*28.96/WF
      PHI=AFST/AFR
      WM1=1.0+4.76*(CA+0.25*HA)/PHI
      W=(P1*V1)/(WM1*RMOL*T1)
C TOTAL MOLES(KGMOL) OF O2, N2, CNHM AT TRAPPED CONDITION
      FAMOL=PHI/(4.7619*SOX)
      AFMOL=1.0/FAMOL
      ATMOL=1.0/(1.0+FAMOL)
      WRITE(6,164) AFST,PHI
  164 FORMAT(1H ,//10X,29HAIR FUEL RATIO STOICHIOMETRIC,F8.3//
     1   10X,20HFUEL AIR EQUIVALENCE,F10.3)
      WRITE(6,166)
  166 FORMAT(1H1,//20X,35HFOLLOWING SPECIES IN PERCENT VOLUME,
     1    4X,37HENERGY (ABSOLUTE) IN J PER KGMOL FUEL//)
      WRITE(6,115)
  115 FORMAT(1H ,3X,4HSTEP,3X,9HVOL(M**3),2X,10HPRESS(BAR),
     11X,7HTEMP(K),4X,3HCO2,6X,2HCO,7X,3HH2O,6X,2HH2,7X,2HO2,7X,
     12HN2,6X,4HCNHM,6X,6HENERGY/)
C CONVERT HEAT OF REACTION FROM J/KG TO J/KGMOL
      QVS=QVS*WF
      DO 4 I=1,4
    4 B(I)=0.0
      B(5)=0.21*ATMOL*WM1
      B(6)=0.79*ATMOL*WM1
      B(7)=1.0/(AFMOL+1.0)*WM1
      P(1,1)=P1
      T(1,1)=T1
      V(1,1)=V1
      EO=0.0
      DO 3 II=1,7
      VAF(II)=B(II)/WM1*100.0
      EO=EO+B(II)*U(II,7)
    3 VAFA(1,II)=VAF(II)
      N=1
      PBAR=P(1,1)*1.0E-5
      PBARA(1)=PBAR
      NC=ND+1
C INTERNAL ENERGY AT TRAPPED STATE FOR REACTANTS
      DO 10 I=5,7
   10 CALL FNU(FU(I),FUDT(I),FV(I),I,U,T1,0)
      CALL FNUT(FUT,FUDTT,WM,5,7,B,FU,FUDT)
      ET1=RMOL*T1*FUT
      ENERGY(1,1)=ET1
      CV=RMOL*FUDTT/WM
C START CALCULATION FOR COMPRESSION STROKE
      DO 14 N=2,NC
C ESTIMATE T2 AND P2
      V2=V1-DV
      T2=T1*(V1/V2)**(RMOL/CV)
      P2=P1*(V1/V2)*(T2/T1)
C INTERNAL ENERGY AT STATE 2
   13 DO 11 I=5,7
   11 CALL FNU(FU(I),FUDT(I),FV(I),I,U,T2,0)
      CALL FNUT(FUT,FUDTT,WM,5,7,B,FU,FUDT)
      ET2=RMOL*T2*FUT
      CV=RMOL*FUDTT/WM
```

```
C WORK
      DW=0.5*(P2+P1)*(V2-V1)
C FIRST LAW
      FE=W*(ET2-ET1)+DW
      ERROR=FE/(W*WM*CV)
      IF(ABS(ERROR).LT.ACCUR1) GO TO 12
C IF FIRST LAW NOT SATISFIED ESTIMATE NEW
C VALUE OF T2 AND P2 AND RETURN TO STATEMENT 13
      T2=T2-ERROR
      P2=P1*(V1/V2)*(T2/T1)
      GO TO 13
C FIRST LAW SATISFIED
C STORE VALUES OF PC,TC,DW AND RESET T1,P1,ET1
   12 T(1,N)=T2
      P(1,N)=P2
      V(1,N)=V2
      ENERGY(1,N)=ET2
      WORK(1,N)=DW
      ET1=ET2
      T1=T2
      P1=P2
      V1=V2
      DO 16 II=1,7
      VAF(II)=B(II)/WM*100.0
   16 VAFA(N,II)=VAF(II)
      WORKT=WORKT+DW
      PBAR=P(1,N)*1.0E-5
      PBARA(N)=PBAR
   14 CONTINUE
C END OF COMPRESSION STROKE
C COMBUSTION AT CONSTANT VOLUME
C SET UP INITIAL CONDITIONS
      TR=T2
      PR=P2
      BB=(2.0/PHI)*(CA+0.25*HA)-(CA+0.5*HA)
C ESTIMATE INITIAL VALUE OF T2
      IF(PHI.LE.1.0) T2=TR+2500.0*PHI
      IF(PHI.GT.1.0) T2=TR+2500.0*PHI-700.0*(PHI-1.0)
C ENERGY LEVEL OF REACTANTS AT TR
      DO 20 I=5,7
   20 CALL FNU(FU(I),FUDT(I),FV(I),I,U,TS,0)
      CALL FNUT(FUTS,FUDTS,WMR,5,7,B,FU,FUDT)
      DO 22 I=5,7
   22 CALL FNU(FU(I),FUDT(I),FV(I),I,U,TR,0)
      CALL FNUT(FUTR,FUDTR,WM,5,7,B,FU,FUDT)
      E1=RMOL*TR*FUTR+EO
C PRINT COMPRESSION STROKE
      DO 24 N=1,NC
      ENERGY(1,N)=ENERGY(1,N)+EO
   24 WRITE(6,159)N,V(1,N),PBARA(N),T(1,N),(VAFA(N,II),II=1,7),
     1ENERGY(1,N)
C PRODUCTS
C SPECIFIC ENERGY LEVEL AND GIBBS FUNCTION
C FOR EACH SPECIES AT T2 AND EQUILIBRIUM CONSTANTS
   60 DO 30 I=1,6
   30 CALL FNU(FU(I),FUDT(I),FV(I),I,U,T2,1)
```

```
      ALKP1=(U(2,1)+U(3,1)-U(4,1)-U(1,1))*(1.0-ALOG(T2))
     1-(FV(2)+FV(3)-FV(4)-FV(1))
     1-(U(2,6)+U(3,6)-U(4,6)-U(1,6))
     1+0.4047E8/(RMOL*T2)
      EKP1=EXP(ALKP1)
      ALKP2=(U(2,1)+0.5*U(5,1)-U(1,1))*
     1(1.0-ALOG(T2))-(FV(2)+0.5*FV(5)-FV(1))
     1-(U(2,6)+0.5*U(5,6)-U(1,6))
     1+2.7969E8/(RMOL*T2)
      EKP2=EXP(ALKP2)
      D=W*RMOL*T2*(EKP2)**2/(PO*V2)
      C=1.0/EKP1-1.0
C SOLVE FOR A (SYMBOL AE)
      CALL CALAE(CA,HA,PHI,EKP1,D,AE)
C SET UP CONSTITUENT MOLS IN PRODUCTS
      A(1)=AE
      A(2)=CA-A(1)
      A(5)=(A(1)/A(2))**2/D
      A(4)=A(1)+2.0*A(5)-BB
      A(3)=0.5*HA-A(4)
      A(6)=3.76*(CA+0.25*HA)/PHI
C SOLVE FOR NEW VALUE OF T2 BY ENERGY BALANCE
      DO 44 I=1,6
   44 CALL FNU(FU(I),FUDT(I),FV(I),I,U,TS,0)
      CALL FNUT(FUTS,FUDTS,WM,1,6,A,FU,FUDT)
      DO 46 I=1,6
   46 CALL FNU(FU(I),FUDT(I),FV(I),I,U,T2,0)
      CALL FNUT(FUTP,FUDTP,WM,1,6,A,FU,FUDT)
      EO=0.0
      DO 182 I=1,6
  182 EO=EO+A(I)*U(I,7)
      E2=RMOL*T2*FUTP+EO
      FE=E2-E1
      CV=RMOL*FUDTP/WM
      ERROR=FE/(WM*CV)
      ER2=ERROR
      IF(ABS(ERROR).LT.ACCUR1) GO TO 50
      T2=T2-ERROR
      GO TO 60
C CONDITIONS AT END OF COMBUSTION
   50 P2=(WM/WMR)*(T2/TR)*PR
      P(2,1)=P2
      T(2,1)=T2
      ENERGY(2,1)=E2
      WORK(2,1)=0.0
      DO 52 II=1,6
   52 VAF(II)=A(II)/WM*100.0
      N=1
      PBAR=P2*1.0E-5
      WRITE(6,160)N,V2,PBAR,T2,(VAF(II),II=1,6),ENERGY(2,1)
C EXPANSION STROKE
      DO 70 N=2,NC
      T1=T2
      P1=P2
      V1=V2
      V2=V1+DV
```

```
C ESTIMATE T2 AND P2
      T2=T1*(V1/V2)**(RMOL/CV)
      P2=P1*(V1/V2)*(T2/T1)
C INTERNAL ENERGY AT STATE POINT 1
      DO 72 I=1,6
   72 CALL FNU(FU(I),FUDT(I),FV(I),I,U,T1,0)
      CALL FNUT(FUT,FUDTT,WM1,1,6,A,FU,FUDT)
      EO=0.0
      DO 74 I=1,6
   74 EO=EO+A(I)*U(I,7)
      E1=RMOL*T1*FUT+EO
C COMPOSITION AT T2
C SPECIFIC INTERNAL ENERGY, GIBBS FUNCTION,
C EQUILIBRIUM CONSTANT
C PRODUCTS
C SPECIFIC ENERGY LEVEL AND GIBBS FUNCTION
C FOR EACH SPECIES AT T2 AND EQUILIBRIUM CONSTANTS
C TEST FOR FREEZING IF YES BYPASS SPECIES CALCULATION
  110 IF(A(4).LT.ACCUR2) GO TO 170
      IF(PHI.LT.1.000.AND.N.GT.NFRST) GO TO 170
      DO 80 I=1,6
   80 CALL FNU(FU(I),FUDT(I),FV(I),I,U,T2,1)
      ALKP1=(U(2,1)+U(3,1)-U(4,1)-U(1,1))*(1.0-ALOG(T2))
     1-(FV(2)+FV(3)-FV(4)-FV(1))
     1-(U(2,6)+U(3,6)-U(4,6)-U(1,6))
     1+0.4047E8/(RMOL*T2)
      EKP1=EXP(ALKP1)
      ALKP2=(U(2,1)+0.5*U(5,1)-U(1,1))*
     1(1.0-ALOG(T2))-(FV(2)+0.5*FV(5)-FV(1))
     1-(U(2,6)+0.5*U(5,6)-U(1,6))
     1+2.7969E8/(RMOL*T2)
      EKP2=EXP(ALKP2)
      D=W*RMOL*T2*(EKP2)**2/(PO*V2)
      C=1.0/EKP1-1.0
C SOLVE FOR A (SYMBOL AE)
      CALL CALAE(CA,HA,PHI,EKP1,D,AE)
C SET UP CONSTITUENT MOLS IN PRODUCTS
  116 A(1)=AE
      A(2)=CA-A(1)
      A(5)=(A(1)/A(2))**2/D
      A(4)=A(1)+2.0*A(5)-BB
      A(3)=0.5*HA-A(4)
      A(6)=3.76*(CA+0.25*HA)/PHI
  170 CONTINUE
C SOLVE FOR NEW VALUE OF T2 BY ENERGY BALANCE
      DO 112 I=1,6
  112 CALL FNU(FU(I),FUDT(I),FV(I),I,U,T2,0)
      CALL FNUT(FUT,FUDTT,WM2,1,6,A,FU,FUDT)
      EO=0.0
      DO 114 I=1,6
  114 EO=EO+A(I)*U(I,7)
      E2=RMOL*T2*FUT+EO
      DW=0.5*(P1+P2)*(V2-V1)
      DE=(E2-E1)*W
      FE=DE+DW
      CV=RMOL*FUDTT/WM2
```

```
      ERROR=FE/(W*WM2*CV)
      IF(ABS(ERROR).LT.ACCUR1) GO TO 200
      T2=T2-ERROR
      P2=P1*(V1/V2)*(WM2/WM1)*(T2/T1)
      GO TO 110
C CONDITION AT END OF VOLUME STEP
  200 P(2,N)=P2
      T(2,N)=T2
      V(2,N)=V2
      ENERGY(2,N)=E2
      DO 75 II=1,6
   75 VAF(II)=A(II)/WM2*100.0
      WORK(2,N)=DW
      WORKT=WORKT+DW
      PBAR=P(2,N)*1.0E-5
      WRITE(6,160)N,V(2,N),PBAR,T(2,N),(VAF(II),II=1,6),
     1ENERGY(2,N)
  159 FORMAT(1H ,3X,I2,2X,E12.4,3X,F8.3,2X,F6.1,2X,7(F7.3,2X),
     1E12.5)
  160 FORMAT(1H ,3X,I2,2X,E12.4,3X,F8.3,2X,F6.1,2X,6(F7.3,2X),
     19X,E12.5)
      XWORKT=WORKT/VS
   70 CONTINUE
      EFFTH=100.0*WORKT/(-QVS*W)
      POWER=WORKT*REV/2000.0
      PMIP=WORKT*1.0E-5/VS
      WRITE(6,162)PMIP,POWER,EFFTH
  162 FORMAT(1H0,14X,12HI.M.E.P(BAR),F10.3,5X,
     120HPOWER(4 STROKE) (KW),
     1F10.3,5X,18HTHERMAL EFFICIENCY,F7.2)
 9999 CONTINUE
      STOP
      END


      SUBROUTINE FNU(F1,F2,F3,I,U,Y,MARK)
      DIMENSION U(7,7)
      F1=0.0
      F2=0.0
      F3=0.0
      DO 1 J=1,5
      S=FLOAT(J)
      F1=F1+U(I,J)*Y**(S-1.0)
      F2=F2+S*U(I,J)*Y**(S-1.0)
      IF(MARK.EQ.0)GO TO 1
      IF(J.EQ.1)GO TO 1
      F3=F3+(U(I,J)*Y**(S-1.0))/(S-1.0)
    1 CONTINUE
      RETURN
      END
```

```
      SUBROUTINE FNUT(F1,F2,F3,I1,I2,X,FU,FUDY)
      DIMENSION FU(7),FUDY(7),X(7)
      F1=0.0
      F2=0.0
      F3=0.0
      DO 1 I=I1,I2
      F1=F1+X(I)*(FU(I)-1.0)
      F2=F2+X(I)*(FUDY(I)-1.0)
    1 F3=F3+X(I)
      RETURN
      END


      SUBROUTINE CALAE(N,M,PHI,KP1,D,AE)
      REAL KP1,M,N
      DIMENSION COEFF(5),AREAL(4),AIMAG(4)
      B=2.0*(N+0.25*M)/PHI-(N+0.5*M)
    2 C=1.0/KP1-1.0
    3 COEFF(1)=2.0*C
      COEFF(2)=4.0*C/D+M-2.0*B*C+2.0*N-4.0*C*N
      COEFF(3)=2.0*C*N**2-4.0*N**2+4.0*B*C*N-3.0*N*M
     1+4.0*N/D-2.0*B*N
      COEFF(4)=2.0*N**3-2.0*B*C*N**2+4.0*B*N**2+3.0*M*N**2
      COEFF(5)=-(2.0*B+M)*N**3
      CALL QUART (COEFF,AREAL,AIMAG)
      AE=-1.0
      DO 1 J=1,4
      IF (ABS(AIMAG(J)).NE.0.0) GO TO 1
      IF (AREAL(J).LT.N.AND.AREAL(J).GT.0.0) GO TO 4
    1 CONTINUE
      IF(AE.GT.0.0.AND.AE.LT.N) RETURN
      WRITE (6,201)
  201 FORMAT (16H AE BEYOND LIMIT)
      STOP
    4 AE=AREAL(J)
      RETURN
      END


      SUBROUTINE QUART (C,XR,XI)
C
C     SOLUTION OF THE QUARTIC EQUATION WITH REAL COEFFICIENTS
C     C(1)*X**4 + C(2)*X**3 + C(3)*X**2 + C(4)*X + C(5) = 0.0
C     USING BROWN'S METHOD
C     THE ROOTS ARE :
C        REAL PART XR(N), IMAGINARY PART XI(N),  N=1,2,3,4
C
      DIMENSION C(5),XR(4),XI(4),AC(4),AQ(3),BQ(3),RT(3)
      EQUIVALENCE (AQ,BQ)
      A3=C(2)/C(1)
      A2=C(3)/C(1)
      A1=C(4)/C(1)
      A0=C(5)/C(1)
```

```
      A=0.5*A3
      AC(1)=1.0
      AC(2)=-A2
      AC(3)=A1*A3-4.0*A0
      AC(4)=A0*(4.0*A2-A3*A3)-A1*A1
      CALL CUBIC (AC,RT,RTI)
      IF (RTI) 5,1,5
    1 IF (RT(1)-RT(2)) 2,3,3
    2 RT(1)=RT(2)
    3 IF (RT(1)-RT(3)) 4,5,5
    4 RT(1)=RT(3)
    5 B=0.5*RT(1)
      IF (B*B-A0) 6,6,7
    6 D=0.0
      CA=SQRT(A*A+2.0*B-A2)
      GO TO 8
    7 D=SQRT(B*B-A0)
      CA=-(0.5*A1-A*B)/D
    8 AQ(1)=1.0
      AQ(2)=A-CA
      AQ(3)=B-D
      CALL QUAD (AQ,XR(1),XR(2),XI(1))
      BQ(2)=A+CA
      BQ(3)=B+D
      CALL QUAD (BQ,XR(3),XR(4),XI(3))
      XI(2)=-XI(1)
      XI(4)=-XI(3)
      RETURN
      END


      SUBROUTINE CUBIC (A,XR,XI)
C
C     SOLUTION OF THE CUBIC EQUATION WITH REAL COEFFICIENTS
C     A(1)*X**3 + A(2)*X**2 + A(3)*X + A(4) = 0.0
C     THE ROOTS ARE :
C       (1) REAL, XR(1)
C       (2) POSSIBLY COMPLEX, REAL PART XR(2), IMAGINARY PART XI
C       (3) POSSIBLY COMPLEX, REAL PART XR(3), IMAGINARY PART -XI
C
      DIMENSION A(4),XR(3),AQ(3)
      IPATH=2
      EX=1.0/3.0
      IF (A(4)) 2,1,2
    1 XR(1)=0.0
      GO TO 16
    2 A2=A(1)*A(1)
      Q=(27.0*A2*A(4)-9.0*A(1)*A(2)*A(3)+2.0*A(2)**3)/
     1(54.0*A2*A(1))
      IF (Q) 4,3,6
    3 Z=0.0
      GO TO 15
    4 Q=-Q
      IPATH=1
    6 P=(3.0*A(1)*A(3)-A(2)*A(2))/(9.0*A2)
```

```
      ARG=P**3+Q**2
      IF (ARG) 7,8,9
    7 Z=-2.0*SQRT(-P)*COS(ATAN(SQRT(-ARG)/Q)/3.0)
      GO TO 13
    8 Z=-2.0*Q**EX
      GO TO 13
    9 SARG=SQRT(ARG)
      IF (P) 10,11,12
   10 Z=-(Q+SARG)**EX-(Q-SARG)**EX
      GO TO 13
   11 Z=-(2.0*Q)**EX
      GO TO 13
   12 Z=(SARG-Q)**EX-(SARG+Q)**EX
   13 GO TO (14,15), IPATH
   14 Z=-Z
   15 XR(1)=(3.0*A(1)*Z-A(2))/(3.0*A(1))
   16 AQ(1)=A(1)
      AQ(2)=A(2)+XR(1)*A(1)
      AQ(3)=A(3)+XR(1)*AQ(2)
      CALL QUAD (AQ,XR(2),XR(3),XI)
      RETURN
      END


      SUBROUTINE QUAD (A,XR1,XR2,XI)
C
C    SOLUTION OF THE QUADRATIC EQUATION WITH REAL COEFFICIENTS
C    A(1)*X**2 + A(2)*X + A(3) = 0.0
C    THE ROOTS ARE :
C      (1) POSSIBLY COMPLEX, REAL PART XR1, IMAGINARY PART XI
C      (2) POSSIBLY COMPLEX, REAL PART XR2, IMAGINARY PART -XI
C
      DIMENSION A(3)
      X1=-A(2)/(2.0*A(1))
      DISC=X1*X1-A(3)/A(1)
      IF (DISC) 1,2,2
    1 X2=SQRT(-DISC)
      XR1=X1
      XR2=X1
      XI=X2
      RETURN
    2 X2=SQRT(DISC)
      XR1=X1+X2
      XR2=X1-X2
      XI=0.0
      RETURN
      END
```

Subject Index